GC/WORKS/2 MOD FORMS & COMMEN~~~~~~~ (1998)

London: The Stationery Office

This document may be cited as - GC/Works/2 Model Forms & Commentary (1998)

© Crown copyright 1998

Published for the Property Advisers to the Civil Estate under licence from the
Controller of Her Majesty's Stationery Office.

Typographical copyright is held jointly by The Stationery Office and the Property Advisers to the
Civil Estate and applications for reproduction should be made in writing to the Copyright Unit,
Her Majesty's Stationery Office, St. Clements House, 2-16 Colegate, Norwich NR3 1BQ

First published 1998

ISBN 0 11 702152 0

CONTENTS

LEGAL BACKGROUND

NOTICES

All parties must rely exclusively upon their own skill and judgement, or upon those of their advisers, when making use of this document. Neither the Crown, nor Pinsent Curtis, nor any other contributor, assumes any liability to anyone for any loss or damage caused by any error or omission, whether such error or omission is the result of negligence or any other cause. Any and all such liability is disclaimed.

The introduction and commentary do not form part of, and shall not affect the interpretation of, any contract.

INTRODUCTION

GC/Works/2 (1998) is a new edition of the standard Government form of contract for minor UK building and civil engineering works, and for demolition works, replacing *General Conditions of Government Contracts for Building and Civil Engineering Minor Works: Form GC/Works/2 (incorporating Amendments 1 to 5) Edition 2 1990 and General Conditions of Contract for Demolition Works: Form C1010 (September 1990)*.

GC/Works/2 (1998) is published in two volumes -

- General Conditions.
- Model Forms & Commentary.

GC/Works/2 (1998) is for use when lump sum tenders are to be invited on the basis of Specification and Drawings only, without Bills of Quantities, with an Employer's option to require the Contractor to submit a schedule of rates in order to value variations ordered. As a guideline, the typical values appropriate to this contract, when used for normal construction works, would be a minimum of £25,000 and a maximum of £200,000. The contract should also be suitable for use in relation to demolition works of any value. However, the choice of contract will in the end be dictated by the circumstances of the project, the inherent difficulties of the Works and the Site, and the overall balance of risks to be covered with regard to such matters as variations and programme changes.

MODEL FORM 1

ABSTRACT OF PARTICULARS AND ADDENDUM

ABSTRACT OF PARTICULARS

Works:

Site:

Condition 1(1) (Definitions, etc.) Employer

The Employer shall be

of

Conditions 1(1) (Definitions, etc.): Project Manager, and 3(1) (Delegations and representatives)

The Project Manager shall be

*of/whose registered office is at

who shall act generally on behalf of the Employer in carrying out those duties described in the Contract, subject to the following excluded matters:

In relation to such excluded matters, the person or persons authorised to act for the Employer are:

*All the CDM Regulations apply/Only Regulations 7 and 13 of the CDM Regulations apply.

Condition 5 (Insurance)

The minimum amount insured in respect of employer's liability referred to in Condition 5(1) shall not be £10,000,000, but shall be £

The percentage for professional fees referred to in Condition 5(2)(a) shall not be 15%, but shall be   %.

The minimum amount insured in respect of public liability referred to in Condition 5(2)(b) shall be £   for any one occurrence or series of occurrences arising out of one event.

The forms of certificates referred to in Condition 5 are appended.

Condition 9 (Defects in Maintenance Periods)

Other than for the services listed below, the Maintenance Period for the Works shall be   months and shall apply from the day after that on which the Works are completed as certified by the PM.

The Maintenance Period for each of the following services, which shall apply from the day after that on which the Works are completed as certified by the PM, shall be:

Service   Period

months.

months.

months.

Condition 10 (Occupier's rules and regulations)

Condition 10 *shall/shall not apply.

*The occupier's rules and regulations are appended.

Condition 15 (Passes)

Condition 15 *shall/shall not apply.

Condition 16 (Photographs)

Condition 16 *shall/shall not apply.

Condition 21 (Commencement and completion)

Period within which Order to Proceed to be given: Days of the acceptance of the tender. In the absence of such notice the Contractor may take possession 14 Days after the acceptance of tender.

The Date for Completion of the Works shall be *   *the last Day of a period of *weeks/months beginning on the day after the day for commencement stated in the Order to Proceed.

Condition 29 (Finance charges)

The rate at which finance charges shall be payable shall be   % over the rate charged during the relevant period by the Bank of England for lending money to the clearing banks.

Condition 37 (Damages for delay)

* Liquidated damages for delay shall be: £   per Day.

* Damages for delay shall be at large.

*Condition 40 (Determination by Contractor)

*The period of suspension referred to in Condition 40(3)(e) shall not be 182 Days, but shall be   Days.

*Condition 41 (Determination following suspension of Works)

*The period of suspension referred to in Condition 41(1) shall not be 182 Days, but shall be   Days.

**Condition 42 (Adjudication)

The adjudicator shall be

of

or, if he is deceased or unwilling or unable to act, or is not or ceases to be independent of the Employer, the Contractor and the PM,

of

or, if he is deceased or unwilling or unable to act, or is not or ceases to be independent of the Employer, the Contractor and the PM; such other person as the Employer and the Contractor choose by mutual agreement in writing or, failing such agreement, such other person as may be chosen by the President or a Vice President of the Chartered Institute of Arbitrators (or, where the Contract is a Scottish contract, by the Chairman or a Vice Chairman of the Chartered Institute of Arbitrators (Arbiters) (Scottish Branch)) at the request of either the Employer or the Contractor.

The prescribed form of adjudicator's appointment is appended.

**Condition 43 (Arbitration and choice of law)

The arbitrator shall be

of

or, if he is deceased or unwilling or unable to act, or is not or ceases to be independent of the Employer, the Contractor and the PM,

of

or, if he is deceased or unwilling or unable to act, or is not or ceases to be independent of the Employer, the Contractor, and the PM; such other person as the Employer and the Contractor choose by mutual agreement in writing or, failing such agreement, such other person as may be chosen by the President or a Vice President of the Chartered Institute of Arbitrators (or, where the Contract is a Scottish contract, by the Chairman or a Vice Chairman of the Chartered Institute of Arbitrators (Arbiters) (Scottish Branch)) at the request of either the Employer or the Contractor.

Condition 47 (Performance bond)

Condition 47 *shall/shall not apply.

*The performance bond shall be in an amount of   % of the Contract Sum, and not 10%.

*The prescribed form of performance bond is appended.

Condition 48 (Parent company guarantee)

Condition 48 *shall/shall not apply.

*The prescribed form of parent company guarantee is appended.

***Supplementary Conditions and Annexes

The following Supplementary Conditions and Annexes (if any) are incorporated into the Conditions of Contract, and shall prevail over the other Conditions of Contract:

*Delete inapplicable items.

**The same adjudicators and arbitrators should be named in all the Employer's contracts relating to the project, whether with contractors, consultants or others.

***It is recommended that any printed Conditions affected by Supplementary Conditions should be amended and initialled by both parties.

ADDENDUM TO ABSTRACT OF PARTICULARS

Schedule of Design Information

Information on items in Condition 28(2)(a) and (b) listed below is not yet available but will be provided by the PM within the periods indicated below. Items not listed will be provided in time to meet the Contractor's reasonable requirements where these have been notified in reasonable time.

Item	**Weeks

* Delete inapplicable items.

** The weeks are counted from the day for commencement stated in the Order to Proceed issued by the Employer (see Condition 21(1) (Commencement and completion)).

MODEL FORM 2

INVITATION TO TENDER AND SCHEDULE OF DRAWINGS

INVITATION TO TENDER

Works:

Site:

1   ('the Employer') invites you to tender, upon the basis of GC/Works/2 General Conditions (1998), for the Works described in the following enclosed documents:

    (a)   Abstract of Particulars and Addendum;

    (b)   Supplementary Conditions and Annexes (if any) referred to in the Abstract of Particulars;

    (c)   Specification;

    (d)   Drawings listed in the attached Schedule of Drawings;

    (e)   Outline Health and Safety Plan; and

    (f)   Other documents as listed below:

2   Your tender should be submitted on the form of Tender and Tender Price Form also enclosed. Any obvious errors in pricing or errors in arithmetic will be dealt with as stated in the form of Tender.

3   You are required to keep your tender confidential and not divulge to anyone, even approximately, what your tender price is or will be. The sole exception to this is information you may have to give to your insurance company, or broker, in order to compile your tender, but you must stress to them that this information is given in strict confidence.

4   You must not make any arrangements with anyone else about whether or not they should tender, or about their or your tender prices or terms and conditions. You may however, obtain any necessary subcontract quotations.

5   No tendering expenses will be reimbursed by the Employer.

6   The Employer does not bind himself to accept the lowest, or any, tender.

7   Your form of Tender should be submitted in a sealed envelope prominently marked:

FORM OF TENDER FOR
WORKS:
SITE:

The envelopes should bear no external indication of the identity of the tenderer.

8   Tenders must be completed and returned by a.m./p.m. on to:

SIGNED by

for and on behalf of the Employer

Tel:

Fax:

Telex:

Date:

SCHEDULE OF DRAWINGS

Drawings prepared by

Discipline

Drawing No. & Revision No. (if any)	Drawing Title	Date

MODEL FORM 3

TENDER AND TENDER PRICE FORM

TENDER

Works:

Site:

To be returned by   a.m./p.m. on   to

of

1   We have examined GC/Works/2 General Conditions (1998), and the following documents:

    (a) Abstract of Particulars and Addendum;

    (b) Supplementary Conditions and Annexes (if any) referred to in the Abstract of Particulars;

    (c) Specification;

    (d) Drawings listed in the Schedule of Drawings;

    (e) Outline Health and Safety Plan (and confirm that we will provide a statement and details of how we plan to implement and develop it, together with details to establish our competence and resources to comply with the requirements and prohibitions imposed upon us relative to health and safety in the execution and/or management of the Works); and

    (f) Other documents as listed below:

2   We enclose for your approval the enclosed documents, which shall be deemed to form part of our tender, listed below:

3   We have obeyed the rules about confidentiality of tenders and will continue to do so as long as they apply.

4   We undertake, within 7 days of being so required by the Employer, to submit to the Employer for his approval a schedule of rates to be used to value variations in the Works. Notwithstanding such approval, we undertake to satisfy the Employer that the prices in the schedule of rates as approved are fair, and should reasonably be used to value variations in the Works.

5   We agree that, should obvious errors in pricing or errors in arithmetic be discovered in any schedules of rates submitted by us during consideration of this offer, we will be afforded the opportunity of confirming or withdrawing our offer/confirming our offer, or of amending it to correct such errors.*

6   Subject to and in accordance with paragraphs 3 to 5 above and the terms and conditions contained or referred to in the documents listed in paragraphs 1 and 2, we offer to execute the Works referred to in the said documents in consideration of payment by the Employer of the sum shown in our accompanying Tender Price Form, which shall be deemed to form part of our tender, plus reimbursement by the Employer of Value Added Tax in accordance with Condition 27 (VAT).

7   (only applicable if Abstract of Particulars states that Condition 47 (Performance bond) shall apply)

      Our surety/sureties will be   Limited/PLC,
      whose registered office is at

8   (only applicable if Abstract of Particulars states that Condition 48 (Parent company guarantee) shall apply)

      Our ultimate holding company (if any) is   Limited/PLC
      (No.   ), whose registered office is at

9   We agree that differences or questions arising out of or relating to the Contract shall be resolved in accordance with Conditions 42 (Adjudication) and 43 (Arbitration and choice of law) of the General Conditions.

SIGNED by

for and on behalf of

Tel:

Fax:

Telex:

Date:

Employer to delete inapplicable item before issuing tender documents.

TENDER PRICE FORM

Works:

Site:

To be returned by   a.m./p.m. on   to

of

The sum referred to in our accompanying form of Tender is   pounds (£   ).

SIGNED by

for and on behalf of

Tel:

Fax:

Telex:

Date:

MODEL FORM 4

INSURANCE DOCUMENTS (CONDITION 5)

NO. 1: CERTIFICATE OF EMPLOYER'S LIABILITY INSURANCE UNDER CONDITION 5(1) OF THE GENERAL CONDITIONS OF CONTRACT FOR BUILDING & CIVIL ENGINEERING MINOR WORKS GC/WORKS/2 (1998)

1   This certificate relates to a contract (the Contract) for the execution of the following Works -

('the Works') made between

('the Employer') and

('the Contractor'), and is furnished to the Employer.

2   Condition 5(1) (Insurance) of the Contract requires the Contractor to effect and maintain employer's liability insurance.

3   Condition 5(4) (Insurance) of the Contract contains further requirements regarding such insurance.

4   We certify that the Contractor has complied with the above requirements by effecting and maintaining insurance as follows -

Insured:

Insurer:

Policy No:

Period of Insurance: from
                          to

Amount Insured: £   , as required by the Contract.

SIGNED by

for and on behalf of

*Contractor's Insurance Broker/Contractor's Insurer

Tel:

Fax:

Telex:

Date:

*Delete inapplicable items.

NO. 2: CERTIFICATE OF CONTRACTOR'S INSURANCE UNDER CONDITION 5(2) OF THE GENERAL CONDITIONS OF CONTRACT FOR BUILDING & CIVIL ENGINEERING MINOR WORKS GC/WORKS/2 (1998)

1   This certificate relates to a contract ('the Contract') for the execution of the following Works -

('the Works') made between

('the Employer') and

('the Contractor'), and is furnished to the Employer.

2   Condition 5(2) (Insurance) requires the Contractor to effect and maintain certain insurance described in that paragraph.

3   Condition 5(4) (Insurance) of the Contract contains further requirements regarding such insurance.

4   We certify that the Contractor has complied with the above requirements by effecting and maintaining insurance as follows -

Construction 'All Risks' (Condition 5(2)(a))

Insured:

Insurers:

Policy No:

Period of Insurance: from
                                        to

Amount Insured in respect of the Works, Thing and professional fees: £

Excess or Deductible: £   per claim.

Public Liability (Condition 5(2)(b))

Insured:

Insurers:

Policy No:

Period of Insurance: from

to

Amount Insured: £   , being at least the minimum sum stated in the Abstract of Particulars included in the Contract, for any one occurrence or series of occurrences arising out of one event.

Excess or Deductible: £   per claim.

SIGNED by

for and on behalf of

*Contractor's Insurance Broker/Contractor's Insurers

Tel:

Fax:

Telex:

Date:

*Delete inapplicable items.

MODEL FORM 5

PERFORMANCE BOND (CONDITION 47)

THIS BOND is made the   day of

BETWEEN:

(1)
[of] OR [whose registered office is at]

('the Contractor');

(2)
[of] OR [whose registered office is at]

('the Guarantor') and

(3)
of

('the Employer', which term shall include its successors and assignees).

WHEREAS by an Agreement ('the Contract') dated as stated in the Schedule and made between the Employer of the one part and the Contractor of the other part, the Contractor undertook the execution of certain works ('the Works') in accordance with the terms and conditions of the Contract.

NOW THIS DEED WITNESSETH as follows:

1   The Guarantor guarantees to, and covenants with, the Employer:

    (a) that the Contractor has discharged, and shall discharge, all the Contractor's obligations (whether existing or future) under or pursuant to the Contract; and

    (b) that, in the event of the Contractor's default in the discharge of any such obligations, the Guarantor shall pay to the Employer the loss and damage thereby caused to the Employer.

2   The maximum aggregate liability of the Guarantor under this Bond shall not exceed the sum set out in the Schedule ('the Guarantee Amount').

3   The Guarantor shall be, and continue to be, liable under this Bond even if the Contract is or becomes not binding on, or unenforceable against, the Contractor, for any reason whatever. No alterations in the Contract, or in the Works, and no extension of time, forbearance or forgiveness, nor any act, matter or thing whatsoever except Expiry (as defined in the Schedule) or an express release by the Employer, shall in any way release or reduce any liability of the Guarantor under this Bond. References to the Contract in this Bond shall include all amendments, variations and additions to it, whether made before or after the date hereof.

4   Whether or not this Bond shall be returned to the Guarantor, the obligations of the Guarantor under this Bond shall be released and discharged absolutely upon Expiry.

5   The Contractor, having requested the execution of this Bond by the Guarantor, concurs in the terms and conditions of this Bond.

6   The Employer shall be entitled to assign or transfer all or any of the Employer's rights under this Bond without the consent of the Guarantor and/or of the Contractor.

7   The proper law of this Bond shall be the same as that of the Contract. Where the proper law of this Bond is Scots law, the parties prorogate the non-exclusive jurisdiction of the Scottish courts.

IN WITNESS whereof the Contractor and the Guarantor have executed this Deed on the date first stated above.

SCHEDULE

Date of the Contract:

Guarantee Amount: 10% of the Contract Sum, namely   pounds (£   ).

Expiry: Expiry occurs on the date when the first of the following events occurs:

(a)   in the event of the Contractor's default in the discharge of any of the Contractor's obligations under or pursuant to the Contract, the Guarantor shall pay to the Employer the loss and damage thereby caused to the Employer, up to the Guarantee Amount; or

(b)   pursuant to Condition 31(5)(b) (Final Account) of the Contract, the Contractor shall have paid the excess therein specified to the Employer, and any amount due from the Contractor to the Employer pursuant to any decision, award or judgement in, or settlement of, any adjudication, arbitration or other proceedings commenced in respect of the Contract before or within sixty (60) days after the end of the Maintenance Period, or where there is more than one Maintenance Period, the end of the last Maintenance Period to expire, has been paid.

[Upon certification of completion of the Works by the Project Manager under Condition 24 (Certifying completion) of the Contract, the Guarantee Amount shall reduce by one-half.]

NOTE: Where the proper law of the above document is Scots law, the format will be subject to alteration to reflect the requirements of Scots law in relation to the execution of a document.

MODEL FORM 6

PARENT COMPANY CONTRACT PERFORMANCE GUARANTEE (CONDITION 48)

THIS AGREEMENT is made the   day of -

BETWEEN:

(1)

[of] OR [whose registered office is at]

 ('the Guarantor'); and

(2)
of

 ('the Employer', which term shall include its successors and assignees).

WHEREAS by an Agreement ('the Contract') dated and made between the Employer of the one part and ('the Contractor') of the other part, the Contractor undertook the execution of certain works ('the Works') in accordance with the terms and conditions of the Contract.

NOW THIS DEED WITNESSETH as follows:

1   The Guarantor hereby absolutely irrevocably and unconditionally guarantees to the Employer the due and punctual performance by the Contractor of all the obligations on the part of the Contractor under or pursuant to the Contract ('the Terms') and (as a separate stipulation and as primary obligor) agrees that if the Contractor shall in any respect commit any breach of or fail to fulfil any of the Terms, then the Guarantor will forthwith perform and fulfil in place of the Contractor each and every Term in respect of which the Contractor has defaulted or which is unfulfilled by the Contractor. The Guarantor shall be liable to the Employer for all losses, damages, expenses, liabilities, claims, costs or proceedings which the Employer may suffer or incur by reason of the said failure or breach.

2   The Guarantor shall be, and continue to be, liable under this Agreement even if the Contract is or becomes not binding on, or unenforceable against, the Contractor, for any reason whatever. No alterations in the Contract, or in the Works, and no extension of time, forbearance or forgiveness, nor any act, matter or thing whatsoever except an express release by the Employer, shall in any way release or reduce any liability of the Guarantor hereunder. References to the Contract in this Agreement shall include all amendments, variations and additions to it, whether made before or after the date hereof.

3   This guarantee shall remain in full force and effect until performance in full of the Terms, notwithstanding:

    (a) the insolvency or liquidation of the Contractor, the Guarantor or any other person;

    (b) any disclaimer of the Contract by a liquidator of the Contractor; and/or any feature of the Contract, the negotiations prior to the Contractor and the Employer entering into the Contract, or the performance of the Contract, making it ineffective or unenforceable.

4   Until the Terms have been unconditionally and irrevocably performed in full the Guarantor shall not by virtue of any performance or payment made by it or otherwise:

(a)   be subrogated to any rights, security or moneys held or received or receivable by the Employer; or

(b)   be entitled to exercise any right of contribution from any co-surety in respect of such performance and liabilities under any other guarantee, security or agreement; or

(c)   exercise any right of set-off or counterclaim against the Contractor or any such co- surety; or

(d)   receive, claim or have the benefit of any payment, distribution, security or indemnity from the Contractor or any such co-surety; or

(e)   unless so directed by the Employer (when the Guarantor will prove, and turn over any realisations to the Employer, in accordance with such directions) claim as a creditor of the Contractor or any such co-surety in competition with the Employer.

5   No delay or omission of the Employer in exercising any right, power or privilege hereunder shall impair such right, power or privilege or be construed as a waiver of such right, power or privilege nor shall any single or partial exercise of any such right, power or privilege preclude any further exercise thereof or the exercise of any other right, power or privilege. The rights and remedies of the Employer herein provided are cumulative and not exclusive of any rights or remedies provided by law.

6   A waiver given or consent granted by the Employer under this guarantee will be effective only if given in writing and then only in the instance and for the purpose for which it is given.

7   (a)   If at any time any one or more of the provisions of this guarantee is or becomes invalid, illegal or unenforceable in any respect under any law, the validity, legality and enforceability of the remaining provisions hereof shall not be in any way affected or impaired thereby.

(b)   As a separate and alternative stipulation the Guarantor unconditionally and irrevocably agrees that any sum expressed to be payable by it or obligation to be performed by it under this guarantee but which is for any reason (whether or not now existing and whether or not now known or becoming known to the Guarantor) not recoverable from or enforceable against the Guarantor on the basis of a guarantee shall nevertheless be recoverable from or enforceable against the Guarantor as if the Guarantor were the sole principal debtor or obligor (where relevant).

8   (a)   Any notice, demand or other communication to be served under this guarantee may be served upon the Guarantor only by posting by first class post or delivering the same or sending the same by telex or facsimile transmission to the Guarantor at its address, or telex or facsimile number shown below:

Address:

Telex:

Fax:

or at such other address or number as the Guarantor may from time to time notify in writing to the Employer.

(b) Any notice, demand or other communication to be served under this guarantee may be served upon the Employer only by posting by first class post or delivering the same or sending the same by telex or facsimile transmission to the Employer at its address, or telex or facsimile number shown below:

Address:

Telex:

Fax:

or at such other address or number as the Employer may from time to time notify in writing to the Guarantor.

9   A notice or demand served by first class post shall be deemed duly served on the second business day after the date of posting and a notice or demand sent by telex or facsimile transmission shall be deemed to have been served at the time of transmission unless served after 5.00 p.m. in the place of intended receipt in which case it will be deemed served at 9.00 a.m. on the following business day. For the purposes of this paragraph business day means a day on which commercial banks are open for business in London.

10   In proving service of any notice it will be sufficient to prove, in the case of a letter, that such letter was properly stamped or franked first class, addressed and placed in the post and, in the case of telex or facsimile transmission, that such telex or facsimile was duly transmitted on a business day to a current telex or facsimile number of the addressee at the address referred to above.

11   The Employer shall be entitled to assign or transfer all or any of the Employer's rights under this guarantee without the consent of the Guarantor.

12   The proper law of this guarantee shall be the same as that of the Contract. Where the proper law of this guarantee is Scots law, the parties prorogate the non-exclusive jurisdiction of the Scottish courts.

IN WITNESS whereof the Guarantor has executed this Deed on the date first stated above.

NOTE: Where the proper law of the above document is Scots law, the format will be subject to alteration to reflect the requirements of Scots law in relation to the execution of a document.

MODEL FORM 7

ADJUDICATOR'S APPOINTMENT (CONDITION 42)

THIS AGREEMENT is made the   day of -

BETWEEN:

(1)
of

('the Employer', which term shall include its successors and assignees);

(2)
[of] OR [whose registered office is at]

('the Contractor'); and

(3)
of

('the Adjudicator').

WHEREAS:

(A)   The Employer has entered into a contract dated   ('the Contract') with the Contractor for the execution of certain Works, and a copy of the Contract has been supplied to the Adjudicator.

(B)   The Adjudicator has agreed to act as [adjudicator] OR [named substitute adjudicator] in accordance with the Contract.

NOW THIS DEED WITNESSETH as follows:

1   The Adjudicator shall, as and when required, act as [adjudicator] OR [named substitute adjudicator] in accordance with the Contract, except when unable so to act because of facts or circumstances beyond his reasonable control.

2   The Adjudicator confirms that he is independent of the Employer, the Contractor, and the Project Manager under the Contract, and undertakes to use reasonable endeavours to remain so, and that he shall exercise his task in an impartial manner. He shall promptly inform the Employer and the Contractor of any facts or circumstances which may cause him to cease to be so independent.

3   The Adjudicator hereby notifies the Employer and the Contractor that he will comply with Condition 42 (Adjudication) of the Contract, and its time limits.

4   The Adjudicator shall be entitled to take independent legal and other professional advice as reasonably necessary in connection with the performance of his duties as adjudicator. The reasonable net cost to the Adjudicator of such advice shall constitute expenses recoverable by the Adjudicator under this Agreement.

5	The Adjudicator shall comply, and shall take all reasonable steps to ensure that any persons advising or aiding him shall comply, with the Official Secrets Act 1989 and, where appropriate, with the provisions of Section 11 of the Atomic Energy Act 1946. Any information concerning the Contract obtained either by the Adjudicator or any person advising or aiding him is confidential, and shall not be used or disclosed by the Adjudicator or any such person except for the purposes of this Agreement.

6	The Employer and the Contractor shall pay the Adjudicator fees, expenses and other sums (if any) in accordance with the Contract and the Schedule, plus applicable Value Added Tax.

7	The Adjudicator is not liable for anything done or omitted in the discharge or purported discharge of his functions as adjudicator, unless the act or omission is in bad faith. Any employee or agent of the Adjudicator is similarly protected from liability.

8	The proper law of this Agreement shall be the same as that of the Contract. Where the proper law of this Agreement is Scots law, the parties prorogate the non-exclusive jurisdiction of the Scottish courts.

IN WITNESS whereof the Employer, the Contractor and the Adjudicator have executed this Deed in triplicate on the date first stated above.

SCHEDULE

Adjudicator's Fees, Expenses, etc..

NOTE: Where the proper law of the above document is Scots law, the format will be subject to alteration to reflect the requirements of Scots law in relation to the execution of a document.

MODEL FORM 8

ORDER TO PROCEED (CONDITION 21)

Employer: [*insert name and address*]
Project: [*insert short description*]
Project No:
Contract No:
Contractor: [*insert name and address*]

To: the Contractor

Date:

The Contractor is hereby required to proceed with the Works on [*date*].

This Order to Proceed is given under Condition 21 (Commencement and completion).

Project Manager

MODEL FORM 9

CERTIFICATE OF COMPLETION (CONDITION 24)

Employer: [*insert name and address*]
Project: [*insert short description*]
Project No:
Contract No:
Contractor: [*insert name and address*]

To: the Contractor

Date:

It is hereby certified under Condition 24 (Certifying completion) that the Works were completed in accordance with the Contract on [*date*].

*This certificate is given without prejudice to the Contractor's obligation to complete the outstanding items listed on the attached schedule.

Project Manager

Delete inapplicable items. The PM is not obliged to issue this certificate subject to a schedule of outstanding items or 'snagging list', but such a concession is often given in practice.

MODEL FORM 10

MAINTENANCE CERTIFICATE (CONDITIONS 9 AND 24)

Employer: [insert name and address]
Project: [insert short description]
Project No:
Contract No:
Contractor: [insert name and address]

To: the Contractor

Date:

It is hereby certified under Condition 24 (Certifying completion) that, the [last*] Maintenance Period having expired, the Contractor has complied with Condition 9 (Defects in Maintenance Periods).

Project Manager

*Delete inapplicable items.

MODEL FORM 11

PROJECT MANAGER'S INSTRUCTION (CONDITION 25)

Employer: [*insert name and address*]
Project: [*insert short description*]
Project No:
Contract No:
Contractor: [*insert name and address*]

To: the Contractor

Date:

The Contractor is hereby instructed under Condition 25 (PM's Instructions) as follows:

[*insert details of Instruction*]

*This Instruction is given in confirmation of an oral Instruction given in emergency under Condition 25(3) on [*date*] to [*insert name of employee or agent of Contractor*].

*Prior to the issue of this Instruction, the Employer and the Contractor have agreed that the lump sum total price of complying with it is   pounds (£   ), excluding VAT. Such agreement shall be confirmed by the Contractor's acknowledgement of receipt of this Instruction.

*The Contractor is requested to submit to the Project Manager a written quotation of the lump sum total price of complying with this Instruction. Pursuant to Condition 25(3), this Instruction is conditional upon agreement of such a lump sum price, pending which agreement the Contractor is not to begin complying with the Instruction.

*The Contractor is requested to submit to the Project Manager a written quotation of the lump sum total price of complying with this Instruction. This Instruction is not conditional upon agreement of such a lump sum price, and the Contractor shall immediately begin complying with the Instruction.

In accordance with Condition 25(3), the Contractor is required immediately to acknowledge receipt of this Instruction.

Project Manager

*Delete inapplicable items.

I acknowledge receipt of the above Instruction on

For and on behalf of the Contractor

MODEL FORM 12

INTERIM PAYMENT CERTIFICATE (CONDITIONS 30 AND 32)

Employer: [*insert name and address*]
Project: [*insert short description*]
Project No:
Contract No:
Contractor: [*insert name and address*]

To: the Employer

Copied to: the Contractor

Date:

It is hereby certified under Condition 32 (Certifying payments) that the net sum not previously certified (taking into account retention and all set-off or abatement to which the Employer is entitled, but exclusive of VAT) to which the Contractor is entitled under Condition 30 (Advances on account) is   pounds (£   ), calculated on the following basis:

Condition 30(2) (Advances on account).   £   £

(a) (i)   97% of the value of the work executed on the Site; and 97% of the value of any Things for incorporation which have been reasonably delivered to the Site and are adequately stored and protected against damage by weather and other causes, but which have not been incorporated in the Works.   - - - - - - - - - - - - - -

   (ii) Less deduction under Condition 30(2)(a)(ii) in respect of Things for incorporation on account of which an advance has been made under Condition 30(2)(a)(i) and which have been incorporated in the Works.   - - - - - - - - - - - - - -
_____

_____   - - - - - - - - - - - - - -

(b) 100% of any amount determined by the PM under Condition 28 (Prolongation and disruption) in respect of the relevant month.

- - - - - - - - - - - - - -

(c) 100% of any amount calculated under Condition 29 (Finance charges). Sub-total of (a) - (c).

- - - - - - - - - - - - - -
- - - - - - - - - - - - - -

(d) Less sum agreed to be credited by the Contractor for old materials.

- - - - - - - - - - - - - -

Certified sum   £_____

_____

Project Manager

This form is intended to comply with the requirements of Part 2 of the Housing Grants, Construction and Regeneration Act 1996.

MODEL FORM 13

FINAL PAYMENT CERTIFICATE (CONDITIONS 31 AND 32)

Employer: [*insert name and address*]
Project: [*insert short description*]
Project No:
Contract No:
Contractor: [*insert name and address*]

To: the Employer

Copied to: the Contractor

Date:

It is hereby certified under Condition 32 (Certifying payments) that the net sum not previously certified (taking into account retention and all set-off or abatement to which the Employer is entitled, but exclusive of VAT) to which the Contractor is entitled under Condition 31(5) (Final Account) is   pounds (£   ), calculated on the following basis:

£

Final Sum

Less amount previously certified

Certified sum   £ _____

Project Manager

MODEL FORM 14

NOTICE OF INTENTION TO WITHHOLD PAYMENT (CONDITION 33)

Employer: [*insert name and address*]
Project: [*insert short description*]
Project No:
Contract No:
Contractor: [*insert name and address*]

To: the Contractor

Date:

Notice is hereby given that the Employer proposes to withhold payment of   pounds (£   ) upon the ground that [*insert details*]

OR Notice is hereby given that the Employer proposes to withhold payment of a total of   pounds (£   ) upon the grounds that [*insert details of each ground for withholding and the amount proposed to be withheld attributable to each ground*]

This notice is given under Condition 33 (Withholding payment).

Project Manager

MODEL FORM 15

NOTICE OF NON-COMPLIANCE WITH INSTRUCTION (CONDITION 36)

Employer: [*insert name and address*]
Project: [*insert short description*]
Project No:
Contract No:
Contractor: [*insert name and address*]

To: the Contractor

Date:

Notice is hereby given that the Employer requires compliance, within [*insert number*] days of the date of this notice, with the following Instruction:

[*insert details*]

This notice is given under Condition 36 (Non-compliance with Instructions).

Project Manager

MODEL FORM 16

EMPLOYER'S NOTICE OF DETERMINATION (CONDITION 38)

Employer: [*insert name and address*]
Project: [*insert short description*]
Project No:
Contract No:
Contractor: [*insert name and address*]

To: the Contractor

Date:

Notice is hereby given that the Contract is hereby determined, upon the ground mentioned in Condition 38(2)*(a)(b)(c)(d)(e)(f)(g)(h) (Determination by Employer), in that [*insert details*]

This notice is given under Condition 38(1) (Determination by Employer).

Project Manager

Delete inapplicable items.

MODEL FORM 17

EMPLOYER'S NOTICE OF INTENTION TO REFER TO ADJUDICATION (CONDITION 42)

Employer: [*insert name and address*]
Project: [*insert short description*]
Project No:
Contract No:
Contractor: [*insert name and address*]

To: the Contractor

Date:

Notice is hereby given that the Employer intends to refer to adjudication the following dispute:

[*insert details*]

This notice is given under Condition 42(1) (Adjudication).

Project Manager

MODEL FORM 18

EMPLOYER'S NOTICE OF REFERRAL TO ADJUDICATION (CONDITION 42)

Adjudicator: [*insert name and address*]
Employer: [*insert name and address*]
Project: [*insert short description*]
Project No:
Contract No:
Contractor: [*insert name and address*]
Project Manager: [*insert name and address*]

To: the Adjudicator

Copied to: the Contractor and the Project Manager

Date:

Notice is hereby given that a dispute, difference or question arising under, out of, or relating to the Contract is hereby referred to you for adjudication.

The matter in dispute is specified in the First Schedule.

The principal facts and arguments relating to the matter in dispute are set out in the Second Schedule.

All relevant documents in the possession of the Employer are specified in the Third Schedule, and copies of such documents are enclosed with this notice.

This notice is given under Condition 42(1) (Adjudication).

For and on behalf of the Employer

[*insert Schedules*]

COMMENTARY

This commentary has been prepared as a general guide to the main points covered by each Condition, but without attempting exhaustive analysis. Some practice and policy context has been included, but this is not the primary aim of the commentary. For further guidance on best practice and policy see, for example:

- Other PACE Central Advice Unit publications such as the Guide to the Appointment of Contractors and Consultants , and periodic Information Notes on current issues.

- Central Unit on Procurement guides, and other advice from the Treasury (the Central Unit on Procurement has recently become the Procurement Practice & Development Group of the Treasury).

- Publications of the Construction Industry Board.

Specific legal advice should be taken if it is proposed to amend any of the General Conditions.

It is important that the appointments of the Employer's construction consultants should appropriately correspond with the Contract, so that they are required to carry out their duties as provided by the Contract.

GENERAL CONDITIONS

The General Conditions are an expanded version of the *General Conditions of Contract for Building, Civil Engineering, Mechanical and Electrical Small Works GC/Works/4 (1998)*. GC/Works/4 (1998) is itself a new edition of the standard Government form of contract for small UK building, civil engineering, mechanical and electrical works, replacing *General Conditions of Government Contracts for Building, Civil Engineering, Mechanical and Electrical Small Works: Form C1001 (September 1990)*.

GC/Works/2 (1998) is an intermediate form of contract, for minor UK building and civil engineering and demolition works, standing between the major works *GC/Works/1 (1998)* forms of contract and the small works *GC/Works/4 (1998)*. Because *GC/Works/2* is for use when lump sum tenders are to be invited on the basis of Specification and Drawings only, without Bills of Quantities, with an Employer's option to require the Contractor to prepare and submit a schedule of rates in order to value variations ordered, its true comparator amongst the three versions of *GC/Works/1 (1998)* is *GC/Works/1 Without Quantities (1998)*. Besides the value and complexity of the relevant works, it is necessary to consider, when choosing between *GC/Works/1 Without Quantities (1998)* and *GC/Works/2 General Conditions (1998)*, whether certain features (such as sectional completion) are required in the relevant Contract.

Features of *GC/Works/1 Without Quantities (1998)* not in *GC/Works/2 General Conditions (1998)*

The following features of *GC/Works/1 Without Quantities (1998)* have no equivalent in *GC/Works/2 General Conditions (1998)*:

Features not in *GC/Works/2 General Conditions (1998)*	Features in *GC/Works/1 Without Quantities (1998)*
Definitions of 'Company', 'Group', 'Employer's Group', 'Contractor's Group', 'Holding Company' and 'Subsidiary'. Therefore, only the actual legal entities respectively constituting the Employer and the Contractor are bound by the Contract. The Employer's Group and the Contractor's Group respectively may *not* be treated as one entity. This does not, of course, affect the possible need to make use of the *GC/Works/2* optional Conditions 47 (Performance bond) and 48 (Parent company guarantee).	1 (Definitions, etc.); 51 (Recovery of sums)
Definition of 'Contract Agreement'. There is no *GC/Works/2* provision for a formal Contract Agreement to be entered into. The Contract will be formed by the Contractor's tender and the Employer's acceptance of the tender, without any formal Contract Agreement. Such a Contract Agreement serves, as a matter of record, in order to avoid later disputes about the identity of the Contract documents. As to the effect of executing a Contract Agreement as a Deed (under English law) or a 'self-proving' document (under Scots law), see below.	1 (Definitions, etc.)
Definitions of 'Milestone', 'Milestone Payment Chart' and 'Stage Payment Chart'. There is no *GC/Works/2* provision for payment of advances on account by reference to a Milestone or Stage Payment Chart. GC/Works/2 Condition 30 (Advances on account) only provides for payment of advances on account by reference to monthly valuation of work executed and Things for incorporation delivered to the Site.	1 (Definitions, etc.); 48 (Advances on account) (Alternatives A (Stage Payment Chart) and B (Milestone Payment Chart)
Definition of 'Programme'. There is no provision under *GC/Works/2* for the Contractor to submit, or the Employer to approve, a programme.	1 (Definitions, etc.); 31(1)(b) (Quality); 33 (Programme)
Definition of 'QS'. There is no provision for a Quantity Surveyor in *GC/Works/2*. Any remaining functions of the QS under GC/Works/2 are assigned to the Project Manager.	1 (Definitions, etc.)

Definition of 'Retention Payment'. There is no provision for payment of advances on account without retention under *GC/Works/2*, under which retention is 3%, rather than 5%.	1 (Definitions, etc.); 48A (Retention payment bond)
Definition of Section'. There is no provision for sectional completion of the Works in *GC/Works/2*.	1 (Definitions, etc.); 34 (Commencement and completion)
Teamworking. There is no provision under *GC/Works/2* for project team meetings, separate from the progress meetings to be held under *GC/Works/2* Condition 22 (Progress meetings).	1A (Fair dealing and teamworking)
Delegations and representatives. There is no provision under *GC/Works/2* for the PM to delegate any of his powers or duties, or to appoint a Clerk of Works or Resident Engineer.	4 (Delegations and representatives)
Contractor's agent and employees.	5 (Contractor's agent); 6 (Contractor's employees)
Insurance. The Contractor is to insure in respect of construction 'all risks' and public liability, usually under his annual policies. There is no equivalent of the more elaborate insurance arrangements set out in Alternatives B and C, mentioned opposite.	8 (Insurance) (Alternatives B and C)
Professional indemnity insurance for design.	8A (Professional indemnity insurance for design)
Setting out.	9 (Setting out)
Design (by Contractor).	10 (Design)
Intellectual property rights.	12 (Intellectual property rights)
Nuisance and pollution.	14 (Nuisance and pollution)
Returns.	15 (Returns)
Foundations.	16 (Foundations)
Covering work.	17 (Covering work)
Measurement.	18 (Measurement)
Programme.	33 (Programme)
Early possession.	37 (Early possession)
Acceleration and cost savings.	38 (Acceleration and cost savings)
Bonuses (for early completion of the Works).	38A (Bonuses)

Valuation of Instructions under *GC/Works/2* Condition 26 (Valuation of Instructions) is much simplified.	41 (Valuation of Instructions - Principles); 42 (Valuation of Variation Instructions); 43 (Valuation of other Instructions)
Payment of advances on account without retention. There is no provision for such payment under *GC/Works/2*.	48A (Retention payment bond)
Mobilisation (or advance) payment. There is no provision for such payment under *GC/Works/2*.	48B (Mobilisation payment)
Payment for Things off-Site. There is no provision for such payment under *GC/Works/2*.	48C (Payment for Things off-Site)
Emergency work.	54 (Emergency work)
Subletting, nomination and insolvency of nominated subcontractors or suppliers. See *GC/Works/2* Condition 44 (Assignment and subletting). In particular, there is no provision in *GC/Works/2* for nomination of subcontractors or suppliers by or on behalf of the Employer.	62 (Subletting); 63 (Nomination); 63A (Insolvency of nominated subcontractors or suppliers)
Collateral warranties. There is no provision under *GC/Works/2* for collateral warranties.	68 (Collateral warranties)

The above table accounts for *GC/Works/1 Without Quantities (1998)* having seventy-seven Conditions (including those which have numbers but are not used), and *GC/Works/2 General Conditions (1998)* only forty-nine.

CONTRACT DOCUMENTATION, INFORMATION AND STAFF

Condition 1 (Definitions, etc.)

This Condition ensures that consistent terminology is used throughout the Contract (including, therefore, the Specification, etc.) by fixing key definitions to be applied whenever the term is printed with an initial capital. In addition, rules are set out for serving formal notices and calculating periods of time under the Contract.

Paragraph (1) sets out the definitions which are to apply in the Contract. Therefore, the definitions will apply in relation to all documents forming part of the Contract, and not only in relation to the Conditions of Contract, as defined.

The definitions are to apply 'unless the context otherwise requires'. Therefore, for example, if a defined term is used in a Contract document, and the relevant context makes it clear that a meaning different from the definition is actually intended, the meaning actually intended is to be used, but only in that specific instance. Clearly, it is not a desirable practice to use a defined term with an intended meaning different from that given in Condition 1(1), as that may cause confusion. In accordance with normal legal practice, defined terms begin with a capital letter.

Most of the definitions are self-explanatory. However, the following further comments may be helpful:

- *'the Accepted Risks'* are risks which are usually uninsurable. The Employer 'accepts' ie assumes, these risks. See, for example, the proviso to Condition 5(2) (Insurance) and Condition 8(5)(b) (Loss or damage).

- *'the Contract'* means the Contract documents mentioned, and no other documents. If documents such as pre-contract correspondence or minutes of meetings are intended to form part of the Contract, they may be incorporated by referring to them appropriately in the Employer's acceptance of the tender. Otherwise, the relevant documents will probably have no contractual effect at all. As to the priority of Contract documents for interpretation purposes, see Condition 2 (Contract documents).

- *'the Contract Sum'* is the original 'sum accepted by the Employer when awarding the Contract'. Naturally, the Contract Sum will usually be adjusted during the course of the Contract - for example, by reason of Conditions 26 (Valuation of Instructions) and 28 (Prolongation and disruption). Such adjustment results in 'the Final Sum'. However, such adjustment will not affect the amount of the performance bond to be given under Condition 47 (Performance bond) (if applicable).

- *'the Contractor'* includes the original Contractor's 'legal personal representatives'. These persons will be the executors or administrators of the estate of a deceased individual Contractor, or of a deceased partner, if the Contractor is a partnership of individuals. At the present time, it will be unusual for a significant construction contract to be awarded to an individual or a partnership of individuals. The vast majority of such contracts will be awarded to companies, which, as bodies corporate, do not have 'legal personal representatives'. The definition also includes 'permitted assignees'. Under Condition 44 (Assignment and subletting), the Contractor is not permitted to assign the Contract without the Employer's consent.

- *'Health and Safety Plan'*. It may be desirable to set a time limit for the Principal Contractor's development of the outline Health and Safety Plan provided to him by the Employer, and to allow the Employer a stated time to examine the developed Plan before construction work starts. Appropriate periods of time will depend upon the complexity of the project, and may be stated, for example, in the Preliminary Section of the Specification.

- '*the Maintenance Period*' (usually of 6 or 12 months from the completion of the Works) does not set a time limit on the Contractor's liability to the Employer for defects in the Works. Such liability continues for the applicable limitation period under English or Northern Ireland law, or for the applicable prescriptive period under Scots law, as to which see below.

- '*Planning Supervisor*' and '*Principal Contractor*'. These definitions make clear that it will not, in every case, be the PM and successful tenderer respectively who fill these positions. This is necessary to ensure maximum flexibility for the Employer and to guard against difficulties in communication during the course of the project when a contractor other than the winning tenderer is serving as Principal Contractor. The Employer may also need to appoint another party as Principal Contractor during the course of the project, for example in cases where the successful tenderer proves incapable of effectively carrying out the duties normally expected of a Principal Contractor.

- '*the PM*'. The Employer may replace the PM at his discretion, either with a direct employee of the Employer, or with an independent consultant.

In paragraph (2), the reference to amendment or re-enactment of legislation will have the effect that if, for example, the Construction (Design and Management) Regulations 1994, or the Prevention of Corruption Acts 1889 to 1916, are amended or re-enacted, they will *automatically* be applied for contractual purposes as so amended or re-enacted, thus avoiding possible future difficulties in the interpretation of the Contract.

The definitions applicable to Sections 104, 105, 106, 107, 108, 112, 113 and 114 of Part II (Construction contracts) of the Housing Grants, Construction and Regeneration Act 1996, are set out within the Act. The above legislation extends to Scotland (see Section 148(2)), but not to Northern Ireland (see Sections 148(1) and (3)). However, corresponding provisions for Northern Ireland have been introduced by an Order in Council pursuant to Section 149 - the Construction Contracts (Northern Ireland) Order 1997.

Sections of the Act correspond to Articles of the Order as follows:

Subject	Section of the Act	Article of the Northern Ireland Order
Construction contracts	104	3
Meaning of construction operations	105	4
Provisions not applicable to contract with residential occupier	106	5
Provisions applicable only to agreements in writing	107	6
Right to refer disputes to adjudication	108	7
Entitlement to stage payments	109	8
Dates for payment	110	9
Notice of intention to withhold payment	111	10
Right to suspend performance for non- payment	112	11
Prohibition of conditional payment provisions	113	12
The Scheme for Construction Contracts	114	13 & 16

With regard to the reckoning of periods of time, the equivalent of Section 116 of the Housing Grants, Construction and Regeneration Act 1996 is Section 39 of the Interpretation Act (Northern Ireland) 1954, but the Sections differ in detail, as reflected in paragraph (4) of the Condition.

Paragraph (3) of the Condition reflects Section 115 of the Housing Grants, Construction and Regeneration Act 1996 and Article 14 of the Construction Contracts (Northern Ireland) Order 1997.

Also under paragraph (3), unless otherwise provided by the Contract, certificates of the PM are to be issued to the Contractor. However, under Condition 32 (Certifying payments), payment certificates are to be issued to the Employer, with a copy to the Contractor.

Paragraph (4) of the Condition reflects Section 116 of the Housing Grants, Construction and Regeneration Act 1996 and Section 39 of the Interpretation Act (Northern Ireland) 1954.

Condition 1A (Fair dealing)

This Condition imposes a mutual duty of fair dealing.

This condition is a 'good faith' provision to reflect *'Constructing the Team'*. It carries some risk because it is a different level of provision from established contractual obligations. It is dependent on parties operating it sensibly, and PACE intends to keep it under review. A general duty is imposed on the parties to 'deal fairly, in good faith and in mutual co-operation, with one another'. All parts of the Contract must be read against the background of this Condition. It will not be sufficient for a party to apply the letter of the Contract, if this would amount to sharp practice or obstructionism. It would be reasonable to expect any such action to count against the responsible party if reviewed by adjudicators and arbitrators in the context of Conditions 42 (Adjudication) and 43 (Arbitration and choice of law). The concept of 'final and conclusive' decisions by or on behalf of the Employer has been dropped, because of its conflict with the principle of universal adjudication of disputes.

Condition 2 (Contract documents)

This Condition deals with the hierarchy of Contract documents, and how discrepancies between them should be resolved.

The 'Contract' is defined in Condition 1(1) (Definitions, etc.), by referring to the relevant Contract documents. See above.

Paragraph (1) deals with possible discrepancies between any Supplementary Conditions and Annexes prescribed by the Abstract of Particulars and the General Conditions - the former will prevail. It also deals with possible discrepancies between the Conditions of Contract and other documents forming part of the Contract. In such cases, the Conditions of Contract will prevail over all other Contract documents. If it is intended to alter the effect of any of the General Conditions, this *must* be effected by Supplementary Condition, and in no other way. Conflicting statements in other Contract documents, such as the Specification, will be overridden by the text of the General Conditions.

Paragraph (2) refers to possible differences between the Specification and the Drawings - the former will prevail, unless the PM instructs otherwise.

Under paragraph (3), figured dimensions on all drawings are preferred to the scale.

Apart from the above, there is no ranking of the Contract documents in order of priority for interpretation purposes, but there are general legal rules of interpretation which will be applied. For example, it is likely that a document prepared especially for the particular Contract will be given priority over a standard printed document, and that a later agreed document will be given priority over an earlier agreed document.

Condition 2, combined with the general legal rules of interpretation, will always provide a contractual answer, whether or not that answer represents the real intention of the parties. However, it is obviously best practice to eliminate all documentary discrepancies which could give rise to disputes.

Condition 3 (Delegations and representatives)

This Condition aims to avoid the problems that can be caused by casual or indiscriminate contacts with the Contractor.

This condition ensures that if the Employer wishes to depart from using the PM as a gateway to the Contractor this must be specifically arranged. Otherwise, all Employer's decisions must be communicated to the Contractor by the PM, even in relation to excluded matters, upon which the PM is not himself entitled to make decisions.

GENERAL OBLIGATIONS

Condition 4 (Conditions affecting Works)

This Condition deals with the Contractor's responsibility to foresee Site conditions, and to price accordingly.

These provisions provide an agreed structure for dealing with Unforeseeable Ground Conditions, as defined. Under the general law, ground conditions, and nearly all other obstacles, would otherwise be at the Contractor's risk.

Paragraph (1) states that the Contractor is 'deemed to have satisfied himself as to all matters and information affecting or likely to affect the execution of, or price tendered for, the Works'. It is, therefore, up to the Contractor to make diligent enquiries concerning the Site and its vicinity, the relevant markets for labour and Things, and any other relevant matters.

Paragraph (2) requires the Contractor to report to the PM alleged Unforeseeable Ground Conditions, as a condition precedent to any right or remedy in respect of such conditions. These cannot be caused by weather, however extreme, but the definition includes artificial obstructions, such as tunnels, pipes or cables. The alleged Unforeseeable Ground Conditions must be conditions which the Contractor 'did not know of, and which he could not reasonably have foreseen having regard to any information which he had or ought reasonably to have ascertained'. This rules out conditions of which the Contractor has been warned by the Employer or by a third party, in the Contract documents or otherwise. It also rules out conditions which are well-known in the vicinity of the Site, or which might reasonably have been foreseen as a result of visual inspection of, or diligent enquiries concerning, the Site. Clearly, because of the nature of the relevant problem, difficult questions of fact may arise under paragraph (2).

The conditions reported under paragraph (2) become Unforeseeable Ground Conditions, as defined in Condition 1(1) (Definitions, etc.), when so certified by the PM under paragraph (3).

Paragraph (4) compensates the Contractor if he 'properly' carries out extra work, or omits work, as a result of Unforeseeable Ground Conditions.

Paragraph (4) also makes it clear that the Contractor will be compensated only in respect of extra or omitted work carried out or omitted *after* the relevant Unforeseeable Ground Conditions have been or should have been certified as such under paragraph (3). Therefore, if the Contractor delays giving notice under paragraph (2), the results for him may be serious.

Paragraph (5) emphasises paragraph (1).

Condition 5 (Insurance)

This Condition details the insurances which must be maintained.

Paragraph (1) requires the Contractor to effect and maintain a minimum level of employer's liability insurance in respect of his employees. The Contractor will be well advised to require in his subcontracts that his subcontractors shall do likewise.

Paragraph (2) provides for the Contractor also to procure construction all risks insurance of the Works, and public liability insurance, in terms which should be suitable (*inter alia*) where the Contractor's existing annual policies are to be used.

Paragraph (3) contains appropriate requirements for the production of evidence of insurance, and paragraph (5) provides for the Employer to insure in default.

Paragraph (4) states certain legal requirements in respect of all insurances under Condition 5. These are intended to safeguard the position of the Employer by requiring that the insurances shall be with reputable insurers operating in the UK, and therefore complying with UK insurance legislation, and that the insurances shall not contain pitfalls, such as very wide exclusions or very high excesses or deductibles. The Third Parties (Rights Against Insurers) Act 1930 and the Third Parties (Rights Against Insurers) Act (Northern Ireland) 1930 are intended to achieve the result that relevant insurance money shall be paid in full to claimants against an insolvent insured, and will not become available for distribution among all the creditors of the insolvent insured, in accordance with insolvency legislation. However, the Acts may be defeated by insurance policy terms to the effect that the insolvent insured must have himself paid the claim before being entitled to recover from the insurers. (See *Firma C-Trade SA v Newcastle Protection and Indemnity Insurance Association (The Fanti)* and *Socony Mobil Oil Co Inc v West of England Ship Owners Mutual Insurance Association (London) Limited (The Padre Island)* [1990] 2 All ER 705 (HL)).

Paragraph (6) deals with existing structures: the risks specified are to be borne by the Employer, subject to the Contractor's liability under Condition 8 (Loss or damage). The Employer may seek to protect himself against the risks specified by normal commercial property insurance. The parties should take professional advice from insurance consultants on insurance matters relating to the Contract and the Works, in particular in order to avoid under-insurance. The insurance position in Northern Ireland differs from that in England, Wales and Scotland in respect of such matters as civil commotion and terrorist damage.

Condition 6 (Statutory notices and CDM Regulations)

This Condition deals with the giving of statutory notices and related matters, and embeds in the Contract the key provisions of the Construction (Design and Management) Regulations 1994.

Paragraphs (1) and (2) require the Contractor to submit notices, etc., required by legislation such as the building regulations, and to pay relevant fees and charges. However, the fees and charges, but not the related overhead or other costs of compliance, are reimbursed to the Contractor under paragraph (3).

With regard to the CDM Regulations, the principle applied in this Condition is that the Contractor accepts the risk of CDM procedures having an adverse effect on the duration and cost of the Works, other than with regard to problems due to the incorrect acts or omissions of the Planning Supervisor.

Paragraphs (4) and (5) ensure that the Contractor, where he is appointed as the Principal Contractor, accepts the risk of the CDM procedures for which he will have primary responsibility.

Paragraph (6) covers instances where a new Principal Contractor is appointed in the course of a project.

Paragraph (7) has been included to ensure that the Contractor provides the Planning Supervisor with the information he needs for the preparation of the health and safety file. If the winning tenderer is not the Principal Contractor, the Planning Supervisor's enquiries will be directed through the Principal Contractor.

Condition 7 (Protection of Works)

This Condition deals with the Contractor s responsibilities for safeguarding the Site and Works.

Paragraph (1) requires the Contractor to be responsible for security and safety on Site.

Paragraph (2) requires the Contractor to comply with regulations governing the storage and use of Things, whether or not the regulations are binding on the Crown.

Condition 8 (Loss or damage)

This Condition spells out the Contractor's liability to make good any loss or damage arising out of, or in any way connected with, the execution of the Works, and complements the insurances effected under Condition 5 (Insurance), which should cover much of the Contractor's liability.

Paragraphs (1) and (6) define very widely the 'loss or damage' to which this Condition applies. The loss or damage is by no means confined to loss or damage to the Works or Things on Site - it will include loss or damage to third party property, personal injuries, and loss of profit.

Paragraphs (2), (3) and (4) require the Contractor to be responsible for all relevant loss or damage, and to indemnify the Employer.

Paragraph (5) reimburses the Contractor 'to the extent that the loss or damage is caused by' the matters stated in subparagraphs (a), (b) and (c). The burden of proof will be on the Contractor to demonstrate this, as he will be trying to prove an exception to the rule, under paragraphs (2) and (3), that he shall pay for all relevant loss or damage. The proviso to subparagraph (5)(c) ensures that subparagraph (5)(c) will *not* apply to loss or damage falling within subparagraph 6(c), namely 'loss or damage to the Works or to any Things on the Site'. Such loss of damage will be dealt with in accordance with Conditions 5 (Insurance), 7 (Protection of Works) and 18(2) and (3) (Vesting).

Condition 9 (Defects in Maintenance Periods)

This Condition deals with defects appearing after completion (as opposed to defects which become apparent during the course of the Works, which are dealt with under Condition 19 (Quality)).

Paragraph (1) requires the Contractor to correct defects in the Works appearing during a Maintenance Period, which the Employer 'considers' to be the Contractor's fault.

Paragraph (2) reimburses the Contractor 'to the extent that any defects were not caused by' the matters stated in subparagraphs (a) or (b). The burden of proof will be on the Contractor to demonstrate this, as he will be trying to prove an exception to the rule, under paragraph (1), that he shall pay for the remedial works.

Paragraph (3) enables the Employer to execute the remedial works at the Contractor's expense, if the Contractor fails to comply with the Condition.

Under paragraph (4), the relevant Maintenance Period begins to run again (for its full term) in respect of the remedial works. This can affect the application of certain other Conditions which refer to the last Maintenance Period to expire - for example, Conditions 5(1) (Insurance), 24 (Certifying completion), and 31(4) and (5) (Final Account).

The Maintenance Period is to be stated in the Abstract of Particulars. Twelve months is normal in UK construction, but six months is also sometimes selected. If a longer period is under consideration, it must be remembered that the Maintenance Period cuts both ways - during that period, the Contractor not only has the *duty*, but the *right*, to correct defects in the Works, and the Employer will not normally be able to employ others to correct defects at the Contractor's expense.

Condition 10 (Occupier's rules and regulations)

This Condition deals with cases where work will occur within the boundaries of official premises.

These optional provisions will very often be required in Government contracts, especially in high-security establishments. The Contractor will be compensated, under Condition 26 (Valuation of Instructions), in respect of Instructions changing the relevant rules and regulations during the execution of the Works.

Condition 11 (Discrimination)

This Condition is included in accordance with assurances given to Parliament concerning race relations, and deals also with unlawful sex discrimination.

Paragraph (1) will have the effect that breaches by the Contractor of the relevant Acts will also be breaches of the Contract.

Paragraph (2) does not impose an absolute obligation. It requires the Contractor to 'take all reasonable steps to ensure' that his employees, agents and subcontractors also observe the Acts.

It is open to Employers to query a tenderer's policy and practice on avoiding unlawful discrimination as part of the prequalification process. This Condition goes further by requiring the Contractor to comply with the relevant legislation.

Condition 12 (Corruption)

This Condition prohibits the Contractor from any corrupt practice, making explicit what would anyway be the legal position under the Prevention of Corruption Acts, which are part of the criminal law. Breach is a ground for determination of the Contract.

Where an improper payment is made in connection with a government contract the onus will be on the accused to show that there was no intent to commit an offence. This condition forbids bribery by the Contractor (paragraph 1) and payment of commission by the Contractor without disclosure to the Employer (paragraph 2).

Paragraph (3) enables the Employer to determine the Contract. Under subparagraph (a), the Employer may determine if 'he is reasonably satisfied' as to breach of the Condition. The Employer is *not* required to wait, for example, while a criminal prosecution is pending, by the conclusion of which the Works would probably be long finished. If the person convicted successfully appeals against his conviction, this does not help the Contractor.

Paragraph (4) enables the Employer to extract from the Contractor the relevant amount.

The Condition refers, at a number of points, to other contracts with the Employer.

Condition 13 (Records)

This Condition reflects the importance of proper record keeping to efficient contract management.

Paragraph (1) requires the Contractor to keep records which it will usually be necessary for him to keep in his own interests. Otherwise, he will be unable to justify relevant claims for payment. Paragraph (2) gives the PM access to those records.

Paragraph (3) requires notice of commencement of daywork, and weekly reporting of daywork.

SECURITY

Conditions 14 (Site admittance), 15 (Passes), 16 (Photographs) and 17 (Official secrets and confidentiality)

These Conditions emphasise the importance of security and confidentiality on Government projects and Sites. Site admittance and pass procedures, where these are warranted, would need to be covered in detail in other contract documents.

Condition 14 is not confined to personnel working on Site.

Conditions 15 and 16 are optional, as they will not be required on all Sites.

In considering whether Condition 16 (Photographs) should be applicable, the effect of its use upon the Contractor's ability to comply with Condition 13 (Records) should be taken into account.

MATERIALS AND WORKMANSHIP

Condition 18 (Vesting)

This Condition operates to transfer title to the Contractor's construction materials and (during the construction period) the Contractor's plant, to the Employer. Its aim is to improve the Employer's position if the Contractor fails to complete the Works, for example due to insolvency.

The Works, being permanently affixed to the Site, will vest in the owner of the Site by operation of law. Things for incorporation will become part of the Works when incorporated in the Works, and will therefore likewise vest in the owner of the Site, no matter who owned them before, or how they were obtained from the true owner.

Paragraph (1) will operate on all Things, including those not for incorporation, but only insofar as the Contractor can pass title to the Employer. The Contractor will frequently not be able to do so, because, for example, the suppliers of Things for incorporation may have validly retained title until payment by the Contractor for such Things; or Things not for incorporation, such as cranes and other constructional plant, may only be hired to the Contractor.

Paragraphs (2) and (3) emphasise the Contractor's risk and responsibility in respect of the Works and Things on Site.

The main effect of paragraph (4) is to prohibit the removal of Things from the Site without the PM's consent.

Condition 19 (Quality)

This Condition sets out the basic obligations accepted by the Contractor in undertaking the Works, and the rights of the PM to reject unsatisfactory work or materials, with provision for testing by an independent expert at the option of the PM (separately from adjudication procedures).

These provisions require the Contractor to undertake the specified Works with reasonable skill and care, which is a concept well understood in law. He is expected to perform to the standard appropriate from an experienced and competent contractor, and the Employer relies on this skill when checking materials and when incorporating them into the Works.

Paragraph (2) is a Contractor's warranty of fitness for purpose of all Things for incorporation, other than those selected by the Employer. This is different from a warranty of fitness for purpose of the Works themselves - which is not provided for by *GC/Works/2*, as the Contractor has no significant responsibility for design. Each individual Thing for incorporation may be fit for its purpose, but the Works may still be unfit for their purposes.

Paragraph (3) requires the Contractor to warn the PM if any Things for incorporation appear, for example, to be unfit for their purposes. The duty to warn will extend, beyond Things for incorporation supplied by or on behalf of the Contractor, to such Things supplied by or on behalf of the Employer - for example, goods and materials issued by the Employer to the Contractor.

Paragraphs (4), (5) and (6) contain a code for:

- demonstration by the Contractor that he is complying with paragraphs (1) and (2);

- wide powers of inspection for the PM, on and off the Site;

- powers of rejection for the PM;

- tests by an independent expert; and

- replacement, rectification or reconstruction of unacceptable Works and Things for incorporation.

Paragraph (7) sets out the Contractor's warranties about his accreditation under any quality control or assurance scheme or system, and how this will be maintained or renewed.

Condition 20 (Excavations)

This Condition deals with the ownership and use of materials or objects, including fossils or antiquities, uncovered by the Contractor during any excavations.

Paragraph (1) makes it clear that material and objects obtained from any work on the Site (including excavations, demolition or dismantling) will belong to the Employer. Some Contracts for new construction may, of course, involve the demolition of pre-existing structures on the Site; and some refurbishment Contracts may involve a significant amount of dismantling. The technical documents included in the Contract should make it clear how the relevant material and objects are to be dealt with.

Paragraph (2) provides for reduction in the Contract Sum, if the Employer provides Things which the Contractor would otherwise have provided.

Paragraphs (3) and (4) deal with the safe handling of fossils, antiquities, etc., and related PM's Instructions, which will be at the Employer's cost.

COMMENCEMENT, DELAYS AND COMPLETION

Condition 21 (Commencement and completion)

This Condition deals with the procedure for translating the specified contract period into actual start and finish dates for the Works.

Paragraph (1) refers to the commencement of the Works pursuant to an Order to Proceed. The effect of the paragraph, and of the related entries in the Abstract of Particulars, is to ensure that the Contractor cannot be left indefinitely waiting to commence the execution of the Works.

The closing part of the paragraph requires the Contractor to complete the Works 'in accordance with the Contract by the Date for Completion'. There is no concept of 'practical' or 'substantial' completion. The PM is not obliged to certify completion under Condition 24 (Certifying completion) subject to a schedule of outstanding items of outstanding work, but this is often done in practice. The certificate of completion included in the Model Forms allows for this.

Obviously, no certificate of completion should be issued while there are outstanding items which significantly impede the beneficial use of the Works by the Employer. Also, if any items on the proposed 'snagging list' are other than trivial, the PM will be well advised, before certifying completion, to extract from the Contractor a written undertaking to deal with those items by specified dates, if necessary with a programme. This should enable the PM to deal more effectively - for example, under Condition 36 (Non-compliance with Instructions) - with subsequent delay by the Contractor in clearing up the items on the 'snagging list'.

Paragraph (2) emphasises that the Contractor must provide all necessary Things, unless the Contract specifically provides otherwise. In this connection, the silence of the Contract cannot benefit the Contractor.

Paragraph (3) enables the Employer, if he so wishes, to recover the Specification(s) and/or Drawing(s).

Paragraph (4) controls débris, etc., arising during the execution of the Works or remaining at the time of completion.

Condition 22 (Progress meetings)

This Condition ensures a regular opportunity for the representatives of all those concerned with progress of the Works to discuss key matters. It sets out a framework for progress meetings and their follow-up.

Regular - usually monthly - progress meetings called by the PM, preceded by a Contractor's report and followed by PM s statements, are important elements in the project management of the Contract. The meetings should ensure that the parties promptly face all relevant issues, especially delay and increased cost, and should also provide valuable contemporaneous evidence in the event of disputes.

Condition 23 (Extensions of time)

This Condition deals with the circumstances under which extensions of time should be awarded to the Contractor, normally following an application by the Contractor but sometimes on the PM's initiative. The fundamental aim of the Condition is to prevent time under the Contract becoming 'at large', and to this extent it operates in the interests of both parties.

The basic contract period stipulated in the Abstract of Particulars should be calculated by the Employer with an allowance for weather conditions, based on such information or assumptions as he has to hand. This may simply be that weather conditions will be normal for the time of year, or will have a neutral effect on a well managed project. The Contractor will need to tender in the light of the period allowed to complete the Works and to accept the risk or advantage that the actual weather will be better or worse than is implied by the period stipulated, subject to subparagraph (b).

The PM may extend time either on the Contractor's request, or on his own initiative, whether or not the Date for Completion has already passed. The Condition sets out the only permitted causes of delay which entitle the Contractor to an extension of time. It is notable that, unlike *GC/Works/1*, unusual weather conditions will be permitted as a cause of delay under subparagraph (b).

The PM should only award extensions of time if the relevant delay is 'due to' one of the permitted causes of delay. Particularly difficult problems may arise in relation to concurrently operating permitted and non- permitted causes of delay. The solution to this problem has been intentionally left to the PM, who must decide, as a matter of fact, whether or not the relevant delay is 'due to' one of the permitted causes of delay.

Condition 24 (Certifying completion)

This Condition deals with the issue by the PM of certificates of satisfactory completion in respect of the Works. This is a separate procedure from payment certification - see Condition 32 (Certifying payments).

The PM's certificate of completion is highly important under several contractual provisions, including Condition 37 (Damages for delay). Completion includes sufficient compliance by the Contractor with Condition 6(7) (Statutory notices and CDM Regulations).

In accordance with the Abstract of Particulars, the Maintenance Period starts to run from the day after the relevant date of completion as certified by the PM. Under Condition 9(4) (Defects in Maintenance Periods), each set of remedial works will have its own Maintenance Period, from the date of making good. After the end of the *last* Maintenance Period to expire, the PM is to issue *one* certificate when the Contractor has complied with Condition 9, and this has an impact on final payment under Condition 31(5) (Final Account).

INSTRUCTIONS AND PAYMENT

Condition 25 (PM's Instructions)

This Condition sets out the wide powers of the PM to issue mandatory Instructions to the Contractor.

The PM's powers under paragraphs (1) and (2) are very wide, especially in view of the closing words of paragraph (2). However, it should be noted that all Instructions will need to satisfy Condition 1A (Fair dealing), and that no Instruction, without the Contractor's consent, may unilaterally alter the terms of the Contract itself. Instructions may vary the Works, but not the Contract.

Under paragraph (3), all Instructions should normally be in writing - they may be handwritten. However, in emergency, Instructions may be given orally and confirmed in writing within 7 Days. As recommended in *'Constructing the Team'* , the PM should normally seek agreement on a lump sum price for an Instruction varying the Works, before issuing it. Where such agreement is reached, the lump sum price should preferably be recorded in the Instruction.

The PM may make his Instruction conditional upon agreement of a lump sum price, pending which the Contractor is not to begin complying with the Instruction.

Condition 26 (Valuation of Instructions)

This Condition sets out the basic rules for the valuation of all Instructions having financial consequences. The preferred method is a pre-agreed lump sum, with other methods applying only in default of such agreement.

Paragraph (1) sets out the bases of valuation. There are four bases, in order of preference, namely, prior quotation and agreement; daywork charges (if prescribed by the Contract); valuation by the PM on the basis of fair rates and prices; and an agreed schedule of rates (not necessarily agreed pre-Contract) in respect of the valuation of variations in the Works.

Under paragraph (2), the valuation in all cases will include for prolongation and disruption whether directly or indirectly stemming from the Instruction, and under paragraph (3) the Contractor is not entitled to payment if the Instruction was necessitated by his default or neglect.

The preferred method of valuation is always prior quotation and agreement. Both parties should approach matters in a reasonable way. Contractors should not load quotes, nor is it expected of those acting for the Employer consistently to reject or recommend against lump sums because they cannot be absolutely sure of their overall price compared to valuation against contractual rates. The lump sum system is inherently less certain and embodies an element of commercial negotiation in order to realise the advantages of pre-pricing.

Condition 27 (VAT)

This Condition sets out how Value Added Tax should be handled under the Contract.

These provisions do not provide for an authenticated receipt procedure in respect of VAT. VAT invoices are required under Condition 32 (Certifying payments).

Paragraph (2) enables the Employer to recover from a defaulting Contractor any extra VAT the Employer has to pay to other contractors employed to make good the default.

Condition 28 (Prolongation and disruption)

This Condition is the general 'claims' provision in the Contract. It is relatively limited due to the exclusion of variations as a source of prolongation and disruption expenses (because these are covered in the specific valuation procedures for Instructions under Condition 26 (Valuation of Instructions)).

Paragraphs (1) and (6) narrowly define 'expense'. The Contractor may only recover out-of-pocket expense under this Condition. Paragraph (1) requires that the Contractor shall have 'properly and directly' incurred the 'expense'. Indirect expense will not do.

The grounds for Contractor's claims under this Condition are set out in subparagraphs (1)(a), (b) and (c), and (2)(a), (b) and (c). Note that, because of the inter-action of subparagraph (1)(b) and paragraph (2), paragraph (2) only relates to *delay* in respect of the matters specified in that paragraph.

Paragraph (3) sets a strict timetable for the Contractor to submit claims in respect of prolongation or disruption. Failure to comply with the timetable will invalidate the relevant claim under this Condition.

Paragraph (4) further narrows the Contractor's opportunity to claim under this Condition. In particular, he may lose his entitlement if he has not served notice under subparagraph (4)(b).

Under paragraph (5), the PM is to value the claim within 28 Days of receiving the Contractor's detailed information under subparagraph (3)(b).

Condition 29 (Finance charges)

This Condition compensates the Contractor in respect of his financial burden if he is incorrectly denied certification.

All payments to the Contractor have to be certified by the PM under Condition 32 (Certifying payments). If certification occurs late then, in accordance with paragraph (1), the Contractor is awarded quarterly compounded interest as set out in paragraphs (2), (3) and (4), subject to the exceptions stated in paragraph (5). Subparagraph (5)(c) is a notable exception in respect of disagreements about the Final Account, but interest may be awarded in such circumstances under Conditions 42 (Adjudication) or 43 (Arbitration and choice of law) - see paragraph (7).

The rate of interest must be fixed in the Abstract of Particulars. It will best be set on a case by case basis, taking account of the likely cost of borrowing to the particular contractor. For larger companies for example a figure of 2% above base could well be appropriate. For smaller companies a higher rate could reasonably apply. It should be noted that proposals for a statutory right to interest for late payment have been drawn up and as this Commentary goes to print the Late Payment of Commercial Debts (Interest) Bill is before Parliament, which if approved will provide for a general default rate of interest to be set by Statutory Instrument. The current proposal is that this should be 8% over base to reflect the usual charge to weaker companies for overdrafts or term lending. However the Bill allows contractual terms to displace the statutory default rate of interest if the contract rate itself provides a 'substantial remedy' within the meaning of the legislation, which envisages consideration to be given to a number of relevant circumstances affecting the contract terms, such as the overall commercial and bargaining balance

Paragraph (6) bars claims for interest or finance charges outside the terms of Condition 29.

Condition 30 (Advances on account)

This Condition deals with the scheme for periodic payment of the Contract Sum to the Contractor.

Paragraph (1) entitles the Contractor to advances on account, subject to certification and invoicing in accordance with Condition 32 (Certifying payments).

In *GC/Works/2*, there is only one payment method - periodic valuation of the work executed and Things for incorporation delivered to the Site. Applications and valuations by the Contractor are required.

Subparagraphs (2)(b) and (c) provide for other additional payments.

Sub-paragraph (2)(d) deals with the situation where the Contractor is allowed to use old materials, in return for a credit in favour of the Employer.

Paragraph (3) enables the Employer to check that the Contractor is duly paying subcontractors and suppliers of Things for incorporation, and to withhold relevant payment if this is not demonstrated.

The Employer's own commitment to paying the main Contractor promptly is of course a demanding one, endorsed by Government policy, good commercial practice and BS 7890: 1996 (*Method for achieving good payment performance in commercial transactions*). Government departments are committed to the CBI Prompt Payers Code, which requires -

- A clear consistent policy for paying bills in accordance with the contract.

- Awareness of this policy at all levels in finance and purchasing departments.

- Agreement of clear payment terms at the outset of a contract, and not seeking to impose unilateral changes on suppliers.

- Providing clear guidance on payment procedures, and a quick system for dealing with complaints and disputes, including advice on which parts of an invoice or claim are disputed.

Paragraph (2) requires the principal element of payments to the Contractor to be subject to deduction of a retention of 3%. Paragraph (4) makes it clear that the Employer continues to hold the entire beneficial interest in the retention - there is no trust or separate bank account in respect of the retention, as provided for by some other published forms of contract.

Condition 31 (Final Account)

This Condition deals with the process and timetable for settling the Final Account within a reasonable time following certified completion of the Works.

Paragraph (1) provides for a further estimated payment to be made to the Contractor as soon as reasonably possible after completion of the Works. This payment is the difference between:

- the Employer's estimate of the Final Sum, less one-half of the accumulated retention; and

- the total amount of advances previously paid under Condition 30 (Advances on account).

This will usually mean (*inter alia*) that the Contractor will receive one-half of the 3% retention.

Under paragraph (2), the PM is to forward the draft final account to the Contractor within 6 months of completion of the Works. If the Final Sum shown in the draft final account is greater than that estimated by the Employer under paragraph (1)(a), the Employer is to adjust the payment made under paragraph (1).

The Contractor has, under paragraph (2), 3 months to react to the draft final account. If he does not appropriately react, he will, under paragraph (3), lose the opportunity to dispute the draft final account.

Paragraph (4) deals with the situation where the Final Sum is determined before the end of the last Maintenance Period to expire. In that case, if the unpaid balance of the Final Sum exceeds the remaining retention (usually $1^{1}/2\%$), the amount in excess of that retention is to be paid to the Contractor. If the Contractor has been paid more than the Final Sum, he is to repay the excess to the Employer.

Paragraph (5) deals with the situation when the PM has certified that the Contractor has complied with Condition 9 (Defects in Maintenance Periods). In that case, if the Final Sum exceeds the amount previously paid to the Contractor, the Employer is to pay the excess to the Contractor, and *vice versa*. If paragraph (4)(a) has already operated, then paragraph (5)(a) will normally have the result that the Contractor will receive the remaining ($1^{1}/2\%$) retention.

Condition 32 (Certifying payments)

This Condition deals with the process of issuing payment certificates, and states the effect they have under the Contract.

Under paragraph (1), *all* net sums due under the Contract are to be certified by the PM to the Employer, with a copy to the Contractor.

Model Form 12, the interim payment certificate, is intended to be used as a presecribed form so as to ensure compliance with Section 110(2) of the Housing Grants, Construction and Regeneration Act 1996, and Condition 33 (Withholding payment).

Paragraph (2) deals with the time for issue of certificates. Contractors often have a preferred date for payment, and therefore provision is made for the first certificate to be issued on a date to be agreed, not later than 28 Days after starting work. If the first certificate is issued on, say, 20 January 1998, subsequent certificates should each be issued on the twentieth day of each month. If there is no sum due, no certificate need be issued.

Under paragraph (3), if the date for certification is not a working day, the certificate is to be issued on the next working day.

Under paragraph (4), in the interests of transparency, the Contractor is immediately to copy the certificates to all his subcontractors and suppliers.

Under paragraph (5), invoicing by the Contractor *follows* certification by the PM. Payment is due 30 Days after invoice.

Paragraph (6) enables the PM, by subsequent certificates, to modify or correct earlier certificates. The 'final certificate for payment' will normally be that which certifies payment under Condition 31(5) (Final Account). The second sentence of the paragraph makes it clear that the Contractor cannot rely upon any certificate of the PM as 'conclusive evidence' of quality or other compliance with the Contract. Of course, the PM's certificates will continue to be evidence in the Contractor's favour, as the PM would not have issued them unless he was satisfied, but the certificates will not prevent the Employer raising later claims in relation to quality, etc., even if a diligent PM'should have detected the problem before issuing the relevant certificate.

Condition 33 (Withholding payment)

This Condition sets out the notice and timing requirements under which all or part of a payment otherwise due under the Contract can be withheld.

These provisions are intended to reflect Sections 110 and 111 of the Housing Grants, Construction and Regeneration Act 1996 (or Articles 9 and 10 of the Construction Contracts (Northern Ireland) Order 1997).

The Employer should obviously be vigilant to comply with this Condition, in any case of doubt, or he will have to make the relevant payment in full. Under paragraph (2), a notice period of 7 Days to withhold payment has been chosen.

The PM's certificates under Condition 32 (Certifying payments) should suffice as notices under paragraph (1) (and Section 110(2) of the Housing Grants, Construction and Regeneration Act 1996), but there is no reason why a separate notice under paragraph (1) should not be served, if necessary.

Condition 34 (Recovery of sums)

This Condition sets out additional rights of set-off for the Employer. The Condition is drawn in wide terms, allowing a global balance to be calculated between the Contractor and the Government under all relevant contracts.

Each company is a separate legal entity. The fact that it may be a member of a group of, say, 100 companies, all owned and controlled by one parent company, makes no difference for contractual purposes. Therefore, for example, if the Employer employs John Doe Construction Limited on a building project, and John Doe Civil Engineering Limited on a civil engineering project, and both companies are members of a group in common ownership, John Doe Construction Limited might become hopelessly insolvent during construction, have its Contract determined, and finally owe the Employer a large sum. However, John Doe Civil Engineering Limited might remain solvent, and go on properly performing its Contract. In that case, the Employer would have to go on paying John Doe Civil Engineering Limited in full, and write off the amount owing from John Doe Construction Limited. Unlike *GC/Works/1, GC/Works/2* does not attempt to deal with this. However, because of the definition of 'the Crown' in Condition 1(1) (Definitions, etc.), the Employer may effect recovery under this Condition in respect of sums due under other contracts with the Crown.

This Condition does not affect the possible need to make use of the optional Conditions 47 (Performance bond) and 48 (Parent company guarantee).

PARTICULAR POWERS AND REMEDIES

Condition 35 (Suspension for non-payment)

This Condition sets out what steps must be taken by the Contractor in order to exercise his right to suspend work for non-payment.

These provisions are intended to reflect Section 112 of the Housing Grants, Construction and Regeneration Act 1996 (or Article 11 of the Construction Contracts (Northern Ireland) Order 1997).

Condition 36 (Non-compliance with Instructions)

This Condition gives the Employer the right to intervene in default of compliance with an Instruction.

The Employer may intervene, without determining the Contract under Condition 38 (Determination by Employer), or otherwise, and charge the Contractor with any *extra* expense.

Condition 37 (Damages for delay)

This Condition deals with how damages for delay are to operate. If liquidated damages for delay are selected, the aim is to pre-set a genuine pre-estimate of loss to the Employer and remove the need to prove actual losses.

Unlike *GC/Works/1*, which provides solely for liquidated damages for delay, paragraph (1) gives alternatives respectively for liquidated damages for delay, and damages 'at large' (to be assessed in the first instance by the PM, subject to subsequent adjudication and arbitration). Damages at large will, broadly speaking, comprise the actual loss or damage suffered by the Employer as a result of the delay. For a public sector Employer to prove such loss or damage in financial terms is frequently difficult or impossible. Therefore, the liquidated damages alternative should normally be used.

The rate of liquidated damages stated in the Abstract of Particulars should be no greater than a genuine '*bona fide*' pre-estimate of the Employer's loss or damage. Otherwise, the rate may be legally attacked and invalidated as an unenforceable penalty. The making of anything better than an intelligent guess may well be difficult or impossible, but the rate chosen should be upheld in the event of a dispute, so long as it is genuine and *bona fide*, and is not a sum plucked out of the air in order to menace the Contractor.

Condition 38 (Determination by Employer)

This Condition sets out the procedures and circumstances under which the Employer may determine the Contract.

Conditions 38-41 set out an extensive code relating to determination of the Contract. However, the opening words of this Condition, and of Conditions 40 (Determination by Contractor) and 41 (Determination following suspension of Works), make it clear that the parties remedies for breach of contract under the general law are preserved. However, because of the high degree of legal uncertainty attached to such remedies, it would be a very unusual case where a party chose to use such remedies in preference to much more precise remedies written into the Contract.

Specific legal advice should be taken before determining any significant Contract, as determination is full of pitfalls, and the source of many legal disputes.

Paragraph (1) enables the Employer to determine the Contract for cause. The Employer's specific grounds for determination are set out in paragraph (2).

As *GC/Works/2* is for minor Contracts, a contractual power for the Employer to determine at will, like that contained in (*inter alia*) *GC/Works/1*, has not been considered necessary.

The references in subparagraphs (2)(c) and (d) to Article 12 of the Construction Contracts (Northern Ireland) Order 1997 (whether or not in force) reflect that, at the time of writing, the Order is not in force, and it is possible that the Contract may be entered into before the Order is in force. In such cases, the definitions of insolvency contained in Article 12 of the Order will nevertheless apply for all purposes of the Contract. If that legislation is amended or re-enacted, it will apply to the Contract as amended or re-enacted, in accordance with Condition 1(2) (Definitions, etc.).

Condition 39 (Consequences of determination by Employer)

This Condition deals with how the cost of determination by the Employer is to be apportioned.

Paragraphs (1)-(3) set out the results of the Employer's determination under Condition 38 (Determination by Employer). Paragraph (1) suspends payments to the Contractor, and allows the Employer to take over the Site, subcontractors and suppliers. When the Works are completed, the PM certifies the cost of completion under paragraph 1(e).

Owing to insolvency law complications, the power to pay subcontractors and suppliers direct under subparagraph (1)(d) should only be exercised with great caution, upon specific legal advice, and should not be exercised if the Contractor is in liquidation or bankruptcy.

Under paragraphs (2) and (3), the Contractor is entitled to credit for all the work he has done, and for the value of Things for incorporation kept by the Employer. The certified cost of completion is then added. If the total is greater than the amount the Contractor would have been paid for completing the Works, the Contractor must pay the excess to the Employer. In the unlikely event that the total is less than the amount the Contractor would have been paid for completing the Works, the Employer must pay the shortfall to the Contractor.

Condition 40 (Determination by Contractor)

This Condition sets out the rights of determination available to the Contractor, and how the consequent cost should be handled.

These provisions introduce an explicit right of determination by the Contractor. In previous editions of *GC/Works/2*, Contractors had to rely on their common law rights to determine in the light of an Employer's acts or omissions. However, the spirit of reciprocal rights and fair play represented by '*Constructing the Team*' makes it appropriate, and sound commercial practice, to ensure that such a key matter is expressly dealt with in the Contract.

Paragraph (1) enables the Contractor to determine the Contract for cause. The Contractor's specific grounds for determination are set out in paragraph (3). Determination under subparagraph 3(a) must be preceded by 30 Days' Contractor's suspension under Condition 35 (Suspension for non-payment).

Subparagraphs (3)(c) and (d) cannot apply while the Employer is a Minister of the Crown, a government department or other Crown agency or authority, because the insolvency procedures specified cannot be applied to the Crown. However, they may well apply if the Crown assigns the Contract under Condition 44 (Assignment and subletting).

Paragraph (5) sets out the results of the Contractor's determination for cause. The items specified in subparagraphs (5)(a)-(d) are totalled - note that (d) includes any Contractor's loss of profit. If the total exceeds all the advances previously paid to the Contractor, the Employer is to pay the excess to the Contractor, and *vice versa*.

Under paragraph (6), the Contractor may remove from Site all Things not for incorporation.

Condition 41 (Determination following suspension of the Works)

This Condition deals with the special circumstances attaching to determination following suspension of the Works.

These provisions enable *either* party to determine the Contract if there is prolonged suspension of the Works, which is not the fault of either party and is not due to weather conditions.

If determination under this Condition occurs, the financial results are as in Condition 40(5) and (6) (Determination by Contractor), but, contrary to Condition 40(5)(d), the Contractor will *not* be entitled to any loss of profit on the Contract.

Condition 42 (Adjudication)

This Condition sets out the procedures to be followed in order to adjudicate disputes, and is intended to avoid festering disputes harming the working relationships on which the project depends for success.

The purpose of adjudication is to provide a form of interim dispute resolution. Time limits remain tight, and comply with the relevant legislation. Adjudication is about speedy and perhaps somewhat rough justice. Some decisions of the Employer, although necessarily open to adjudication, cannot in practice be reversed and any adjustments must therefore be made by way of financial compensation should the adjudicator find that the Employer is at fault. Aside from this, the adjudicator has full powers to correct errors and mistakes made by either side, which have led to disputes.

The Condition reflects Section 108 of the Housing Grants, Construction and Regeneration Act 1996 (or Article 7 of the Construction Contracts (Northern Ireland) Order 1997).

The adjudicator and named substitute adjudicator should, wherever possible, be appointed before, or simultaneously with, entry into the Contract.

In order to avoid diverse decisions by different adjudicators, the same adjudicators should be named in all the Employer's contracts relating to the project, whether with contractors, consultants or others. In some cases, it may also be appropriate for the main contract to specify that the subcontract terms should name the main contract adjudicator as also acting in subcontract disputes.

The adjudicator's powers under paragraph (6) to award simple or compound interest are based on Section 49 of the Arbitration Act 1996.

The progress of a dispute through adjudication could be as follows:

- Either party may give notice at any time of his intention to refer a dispute to adjudication - a form of notice is given in the Model Forms.

- Within 7 Days of that notice, he may by notice of referral actually refer the dispute to the adjudicator - a form of notice of referral is given in the Model Forms.

- The notice of referral must specify the matter in dispute and the principal facts and arguments relating to it; and enclose copies of all relevant documents in the possession of the referring party. Therefore, it may involve quite a considerable mass of documentation, requiring skilful legal drafting and assembly. The task will practically combine the steps which, in an arbitration, would constitute the claimant s points of claim and discovery of documents. Two points are notable in relation to this:

- The referring party may have considerable leisure during which to prepare his notice of referral and accompanying documents, while the other party will only have a strictly limited time to respond.

- The referring party, at least during the construction period, is much more likely to be the Contractor than the Employer, because disputes during that period will often involve the Contractor desiring more money from the Employer, against the wishes of the Employer and his consultants.

- The other party and the PM may submit representations to the adjudicator not later than 7 Days from the receipt of the notice of referral. Naturally, in this limited time, no great elaboration or documentation may be expected. There is nothing effectively to prevent the other party from subsequently submitting further representations and documents to the adjudicator.

- The adjudicator is not required to hold a hearing, although he may do so if he wishes. He has great discretion over his methods of proceeding.

- The adjudicator, under Section 108 of the Housing Grants, Construction and Regeneration Act 1996, is given a strictly limited time to reach his decision. This is why the Condition gives the non-referring party such a limited time to respond to the referring party's notice of referral. Extensive reasoning within the decision may not be possible but a party may seek reasons within 14 days.

- The adjudicator's timetable for decision is derived from Sections 108(2)(c) and (d) of the Housing Grants, Construction and Regeneration Act 1996. He may only extend the time limit for decision (of 28 Days from notice of referral) with the agreement of the Employer and the Contractor, or (by no longer than 14 Days) with the consent of the referring party. However, it is perhaps unlikely that a sensible party will refuse any reasonable request by the adjudicator for an extension of time for the making of the adjudicator's decision. If the party refuses, the next thing he may receive from the adjudicator is a decision adverse to him. If the referring party refuses an extension of time, the adjudicator might well decide that, as he cannot properly evaluate the non-referring party's case within the time allowed, he will throw out the referring party's claim. If the non-referring party refuses an extension of time, the adjudicator might well decide that, as the non-referring party's case might look rather threadbare at that time, he will allow the referring party's claim.

- If the adjudicator issues his decision out of time, it is nevertheless valid.

- The adjudicator has full power to award costs and expenses, just like an arbitrator or judge. This is a very important power, particularly in order to punish unmeritorious or repeated referrals.

- If the adjudicator's decision is that one party shall pay to the other £10,000,000, plus costs, the successful party may immediately enforce that decision by summary judgement and enforcement by the courts. If the court is satisfied that no defence apllies which might warrant trial of the case, a quick judgement is normally available.

Condition 43 (Arbitration and choice of law)

This Condition deals with the procedures to arbitrate disputes, and is intended to operate as a second level of dispute resolution for major issues following completion of the Works.

In order to avoid diverse decisions by different arbitrators, the same arbitrators should be named in all the Employer's contracts relating to the project, whether with contractors, consultants or others.

In the majority of cases, specific legal advice should be taken before beginning, and throughout, any arbitration.

Arbitration is the subject of a large body of statute and case law, developed over very many years. The Arbitration Act 1996, a new statutory code for arbitration in England, Wales and Northern Ireland, has been recently enacted. Following the Dervaid Report (Report to the Lord Advocate on Legislation for Domestic Arbitration in Scotland, 1996, by the Scottish Advisory Committee on Arbitration Law), an Arbitration (Scotland) Bill has been published and is expected to become law in the near future.

Arbitration is a form of private dispute resolution, in which:

- the parties, or an appointing institution, appoint the 'judge' , rather than having him provided by the State; and

- the parties pay, not only their legal costs and expenses, but also the fees of the arbitrator and the cost of accommodation and facilities for hearings.

Arbitration has a certain inherent disadvantage in relation to multi-party disputes, so common in construction projects, and the legal background to this problem has recently been changed by the Arbitration Act 1996. See the later commentary on multi-party disputes and the Arbitration Act 1996, where it is concluded that the Employer may protect himself from the relevant problems by ensuring:

- that all his contracts with contractors, consultants and others relating to the same project provide for the arbitration of disputes;

- that the same arbitrator is appointed under all the contracts, by naming him in each; and

- that the arbitrator has power to order that arbitral proceedings under each of the contracts shall be 'consolidated' (proceed together) and that concurrent hearings shall be held (as contemplated by Section 35 of the Arbitration Act 1996) and, in Scotland, that the arbiter has the same power as the court in respect of the joining of one or more defenders or joining co-defenders or third parties.

Arbitration does have the substantial advantage of being private - so, for example, allowing Employers to keep confidential the prices which they are paying for the relevant construction work, and Contractors their profit margins. Mainly for this reason, arbitration clauses appear in the vast majority of significant construction contracts.

Privacy may be jeopardised, at least to some extent, by recourse to the courts. This may arise, for example, from:

- an application to the court to determine a point of law *during* the arbitration proceedings (under Section 45 of the Arbitration Act 1996); or

- an appeal to the court on a question of law *arising out of* the arbitrator's award (under Section 69 of the Arbitration Act 1996, or under Section 3 of the Administration of Justice (Scotland) Act 1972, or by means of judicial review).

The above rights of recourse to the courts may be excluded by agreement, as expressly permitted by Sections 45 and 69, but this Condition does not do so. It is considered reasonable in principle that the parties should have rights of recourse to the courts on points of law (*not* questions of fact). However, such rights could be excluded in exceptional cases by Supplementary Condition.

Under paragraph (1), all disputes are referable to the arbitrator, except:

- disputes about the 'enforceability or enforcement' of adjudicator's decisions (parties may apply *directly* to the courts to enforce these); and

- VAT or statutory tax deduction scheme disputes, for which statutory adjudication machinery is available.

There are no longer any 'final and conclusive' decisions which cannot be adjudicated or arbitrated.

The arbitrator will have the additional powers listed in paragraph (1). Subparagraphs (d) and (e) will be particularly important in multi-party disputes.

Under paragraph (1), disputes 'shall after notice by either party to the other be referred to the… arbitrator…'. If a party wishes to arbitrate an adjudicator's decision, he must give to the other party the preliminary notice referred to within 56 Days of the adjudicator's decision, or that decision will become unchallengeable.

It should be noted that the preliminary notice to the other party, and the actual reference to the arbitrator, are different steps. Because of subparagraph (2)(a), the preliminary notice may precede the reference by a very considerable time.

Under subparagraph (2)(a), no reference to the arbitrator may be made until after the completion, alleged completion or abandonment of the Works, or the determination of the Contract; and under subparagraph (2)(b), no reference to the arbitrator may be made in respect of a dispute in due course of adjudication.

Subparagraphs (c)-(e) require an immediate 'preliminary meeting', and lay down a timetable for the exchange of formal claims, defences and counterclaims ('pleadings'); for the discovery (disclosure) and inspection of documents in the control of each party; and for a hearing ('if necessary', but it *will* be necessary in any significant case). The target period of 6 months is expeditious by arbitration or litigation standards, bearing in mind that cases which proceed to arbitration are likely to be of some difficulty and complexity, with voluminous papers. Simple, crystal-clear cases are hardly likely to be referred to the arbitrator in the first place, and still less likely to proceed far. Arbitration hearings sometimes last months, after all the pleadings, discovery and so forth have been completed. The arbitrator is meant to give his award within a further 3 months after the hearing.

ASSIGNMENT, SUBLETTING, SUBCONTRACTING AND OTHER WORKS

Condition 44 (Assignment and subletting)

This Condition deals with the assignment of the benefit of the Contract, and the subcontracting of any part of the Works.

Under paragraph (1), the Contractor is *not* permitted to assign the benefit of the Contract, without the Employer's consent, but under paragraph (2) the Employer *is* permitted to assign the benefit of the Contract, without the Contractor's consent.

Neither party is entitled to assign the burden of the Contract - that is, the obligation to perform the Contract.

Assignment of the benefit of the Contract by either the Contractor or the Employer will *not* release the original Contractor or Employer from his present or future liability under the Contract. For example, if the original Employer assigns the benefit of the Contract, and the assignee defaults in payment at *any* time after assignment, the original Employer will still be liable to the Contractor in respect of payments due under the Contract.

Assignment by the Employer transfers to the assignee all the Employer's rights in respect of breach of the Contract by the Contractor, including those in respect of latent and inherent or other defects in the Works. Therefore, it is important to a purchaser of the Site, or anyone else acquiring an interest in the Works, to take such an assignment from the Employer, and assignments of all related bonds, parent company guarantees, etc..

Under paragraph (1), the Contractor is also not permitted to sublet or subcontract, without the PM's consent.

Condition 45 (Provisional sums)

This Condition sets out procedures to be followed when the Employer has included provisional sums in respect of work which cannot be foreseen or costed accurately when tenders are invited.

Provisional sums are only to be expended on PM's Instructions. The Contractor's entitlement to payment is then akin to that in respect of Instructions. The original provisional sum, often a round figure, is of little importance. However, the original provisional sums will be included in the original Contract Sum, and will therefore affect the amount of the performance bond referred to in Condition 47 (Performance bond).

Condition 46 (Other works)

This Condition deals with situations where the Employer must undertake other works on the Site.

Paragraph (1) recognises that, on many Sites, the Employer will be contemporaneously having the Works executed through the main Contract with the Contractor, and also having other works executed by other contractors and suppliers contracted directly to the Employer. The Contractor must give 'reasonable facilities' for those other works - for example, access to the Site and allowing the other contractors and suppliers to use his scaffolding. Clearly 'reasonable facilities' is not a precise expression, and it will be desirable, if the work of the other contractors and suppliers is likely to be significant, to give more details in the other Contract documents - for example, about the supply by the Contractor of utilities, such as water and electricity, and payment for such services.

Paragraph (2) fits in with Condition 8 (Loss or damage).

PERFORMANCE BOND AND PARENT COMPANY GUARANTEE

Conditions 47 (Performance bond) and 48 (Parent company guarantee)

These Conditions enable the Employer to require a performance bond and/or a parent company guarantee in respect of the Contractor's due performance of the Contract.

Performance bonds and parent company guarantees are optional. A performance bond may reasonably be foregone if the Contractor's evident financial strength (and/or that of his parent company guarantor) is so great that insolvency is extremely unlikely. A parent company guarantee may reasonably be foregone if the Contractor's own evident financial strength is so great that his insolvency is extremely unlikely - however, it should also be borne in mind that parent companies have ways of extracting assets from subsidiaries, especially given time. Employers should, therefore, be extremely wary of foregoing parent company guarantees in respect of significant Contracts.

Contractors sometimes argue that, if the Employer has a performance bond, he does not need a parent company guarantee. This is not correct. Performance bonds are always strictly limited in amount (usually 10% of the Contract Sum) and duration, whereas parent company guarantees are usually not. This can be very important if a latent and inherent defect in the Works appears some considerable time after completion.

As in relation to Condition 44 (Assignment and subletting), it is important to a purchaser of the Site, or anyone else acquiring an interest in the Works, to take assignments of all related bonds and parent company guarantees.

Attention is drawn to the Department of the Environment report *'The Use of Performance Bonds in Government Construction Contracts'* , November 1996. Some of the report's recommendations may be summarised as follows:

- 'On demand bonds' should not form part of Government construction contracts, other than in exceptional limited cases, such as advance payment bonds or retention bonds.

- Government departments should, in general, rely on effective pre-qualification and vetting procedures to choose tenderers, and not employ bonds routinely.

- Where bonds are used they should be conditional, and reflect a case-by-case assessment of risks for the particular contract.

- Government departments should take account of the fact that requiring a bond can adversely affect competition for a contract, and that not requiring a bond might give better value for money.

- The use of parent company guarantees should be considered where security is needed against post- completion liabilities.

MODEL FORMS

Specific legal advice should be taken if it is proposed to amend any of the Model Forms.

Model Form 1: Abstract of Particulars and Addendum

This document identifies all the variables which need to be included in the Contract. If in doubt about an entry relating to a particular Condition, reference to the commentary on the General Conditions may be helpful.

The Abstract identifies the Planning Supervisor, and also states whether all the CDM Regulations apply or only 7 and 13. See Regulation 3 (Application of regulations), paragraphs (2), (3) and (8). Where *GC/Works/2 (1998)* is suitable for use, it is most unlikely that only Regulations 7 and 13 of the CDM Regulations will apply.

With regard to Supplementary Conditions and Annexes, see the commentary on Condition 2 (Contract documents).

Model Form 2: Invitation to Tender and Schedule of Drawings

The European Union public procurement rules may require further information to be added to the invitation to tender. See the commentary in relation to those rules, below.

Model Form 3: Tender and Tender Price Form

The Contractor's tender, and the Employer's acceptance thereof, will constitute a contract. The Employer's acceptance of the tender will be a Contract document, under the definition of 'the Contract' in Condition 1(1) (Definitions, etc.), and see the commentary thereon.

Paragraph 2 of the tender will show the documents which the Contractor submits - which will usually be the technical and other documents explaining and elaborating on the tender.

With regard to obvious pricing or arithmetical errors, paragraph 5 contains alternative procedures, one of which should be selected by the Employer before issuing the tender documents. The alternatives are for the tenderer to confirm or withdraw his offer: or to confirm or correct his offer. Guidance on this matter is given in the *PACE 'Guide to the Appointment of Contractors and Consultants'*. Reference may also be made to the National Joint Consultative Committee for Building *'Code of Procedure for Single Stage Selective Tendering'*.

Paragraph 6 provides for the tender price to be submitted separately if the Employer wishes to operate the two envelope tendering process.

Model Form 4: Insurance Documents

These comprise two documents to be given on behalf of the Contractor, namely:

- a certificate of employer's liability insurance under Condition 5(1) (Insurance); and

- a certificate of construction 'all risks' and public liability insurance under Condition 5(2) (Insurance).

See also the commentary on Condition 5 (Insurance).

Model Form 5: Performance Bond

Perhaps the most vital part of the bond is the definition of 'Expiry' in the Schedule. If the project follows the normal course, and the Contractor completes the Works in the usual way, the first part of paragraph (b) will apply, upon discharge of the Final Account. Under the second part of paragraph (b), if any adjudication, arbitration or other proceedings are commenced (by the Employer or the Contractor) in respect of the Contract within 60 days of the expiration of the last Maintenance Period, the bond will remain in force until the proceedings are concluded, and any amount due to the Employer has been paid. If the proceedings were commenced by the Contractor, he may at that time owe substantial legal costs to the Employer.

If determination of the Contract occurs, paragraph (a) of the definition will be applicable.

The optional concluding part of the definition provides for automatic reduction of the bond on completion of the Works, *even in respect of Employer's claims against the Contractor then pending*. For example, the Contractor may owe the Employer damages for delay under Condition 37 (Damages for delay).

Clause 3 is essential, as under the general law even minor changes in the Contract agreed by the Employer and the Contractor may otherwise invalidate the bond.

See also the commentary on Condition 47 (Performance bond).

Model Form 6: Parent Company Contract Performance Guarantee

Note that, unlike the performance bond, this guarantee is not limited in time or amount.

See also the commentary on Condition 48 (Parent company guarantee).

Model Form 7: Adjudicator's Appointment

This document will serve to appoint the adjudicator and substitute adjudicator named in the Abstract of Particulars, whose duties and powers are described in Condition 42 (Adjudication).

The form is drafted as a joint appointment by the Employer and the Contractor.

Model Forms 8-18: Administrative documents

These suggested model administrative forms require no particular comment.

Model Form 12, the interim payment certificate, is intended to comply with Section 110(2) of the Housing Grants, Construction and Regeneration Act 1996, and Condition 33 (Withholding payment), and is therefore more elaborate than most pre-existing forms.

It should be noted that, under Condition 32 (Certifying payments), only the PM signs certificates.

LEGAL BACKGROUND

European Union Public Procurement Rules

As *GC/Works/2 (1998)* is intended primarily for use by Ministers of the Crown, government departments or other Crown agencies or authorities, as Employers, the EU public procurement rules will frequently be relevant and applicable if the value of the proposed Contract is above the specified threshold value. The rules require that certain detailed procedures, including advertising and notice, shall be followed, and this may vary depending upon which of the alternative procedures available are used. Such compliance may well require further information to be added to the invitation to tender.

English, Scots and Northern Ireland law

GC/Works/2 (1998) is suitable for use where the governing or 'proper' law of the Contract is English, Scots or Northern Ireland law. The proper law clauses of the ancillary documents follow the proper law of the Contract itself. The proper law of the Contract depends on the location of the Works. See Condition 43(3) (Arbitration and choice of law). If any other law is to govern the Contract, or affect any contractual obligations, professional advice should be taken in connection with the relevant jurisdiction.

Limitation Periods under English and Northern Ireland law

GC/Works/2 (1998) does not provide for a formal Contract Agreement, whether or not to be executed as a Deed. However, certain ancillary documents, such as performance bonds, parent company guarantees and adjudicators' appointments, should be executed as Deeds, for technical legal reasons relating to the doctrine of consideration. The execution of a document as a Deed also results, in English and Northern Ireland law, in a 12 year, rather than a 6 year, limitation period for claims - for example, Employer's claims in relation to defects in the Works.

Prescriptive Periods under Scots law

Scots law, on the other hand, does not recognise a distinction between Deeds and other written documents creating contractual obligations. Under Scots law (subject to certain exceptions), claims for payment under a contract or for damages for breach of contract will prescribe and become barred 5 years after the obligation became enforceable, subject to the concept of discoverability and to a long-stop of 20 years. The Requirements of Writing (Scotland) Act 1995 distinguishes between documents which are self-proving and those which are not - generally, to be self-proving the signature to the document must be witnessed - but the distinction has no relevance to the duration of the prescriptive period, after which claims will be barred.

Execution of Deeds and other documents under English law

Under English law, an appropriate attestation clause for execution of a Deed by an individual is as follows:

SIGNED as a DEED by
JOHN DOE in the
presence of: (Signed) JOHN DOE

(Signed) RICHARD ROE

of:

A witness to a Deed should not be the spouse or a relative of the signatory. He should add his or her name, address and occupation or description, as illustrated.

If a partnership is executing a Deed, all the partners should sign as above, or an authorised partner or authorised partners may sign on behalf of the partnership as follows:

SIGNED as a DEED by
JOHN DOE for and on behalf of the
(JOHN DOE PARTNERSHIP
in the presence of: Signed) JOHN DOE

(Signed) RICHARD ROE

of:

Unless all partners sign, those signing should prove their authority to execute Deeds on behalf of the partnership, e.g., by a power of attorney or certified extract from their partnership agreement.

A company executing a Deed using a common seal could execute as follows:

THE COMMON SEAL of
JOHN DOE [LIMITED]
OR [PLC] was affixed to
this DEED in the presence of: [COMMON SEAL]

(Signed) JOHN DOE Director

(Signed) JANE JONES Secretary

A company not using a common seal could execute as follows:

SIGNED as a DEED for
and on behalf of
JOHN DOE [LIMITED] OR
[PLC] by:

(Signed) JOHN DOE Director

(Signed) JANE JONES Secretary

The above attestation clauses conform with English law on the execution of Deeds as amended with effect from 31 July 1990 by Section 1 of the Law of Property (Miscellaneous Provisions) Act 1989 and Section 130(2) of the Companies Act 1989. The principal changes from the previous practice are that:

•   if a document is to be a Deed it must be clear *on its face* that it is intended to be a Deed;

•   sealing by individuals (including partners) has been abolished; and

•   a company may choose whether or not to use a common seal.

Attestation clauses should preferably be added at the very end of the relevant document, i.e. *after* any annexations or schedules.

If, exceptionally, a document is executed otherwise than as a Deed, all references to Deed in the relevant text should be replaced by Agreement , and the attestation clauses should be as follows:

SIGNED by JOHN DOE
in the presence of:

OR

SIGNED by JOHN DOE
for and on behalf of the
JOHN DOE PARTNERSHIP
in the presence of:

OR

SIGNED by JOHN DOE
for and on behalf of
JOHN DOE [LIMITED]
OR [PLC] in the presence of:

Execution of Deeds and other documents under Northern Ireland law

Under Northern Ireland law, an appropriate attestation clause for execution of a Deed by an individual is as follows:

SIGNED SEALED and
DELIVERED by JOHN DOE in
the presence of: (Signed) JOHN DOE [SEAL]

(Signed) RICHARD ROE

of:

A witness to a Deed should not be the spouse or a relative of the signatory. He should add his or her name, address and occupation or description, as illustrated.

If a partnership is executing a Deed, all the partners should sign as above.

A company executing a Deed using a common seal could execute as follows:

THE COMMON SEAL of
JOHN DOE [LIMITED]
OR [PLC] was affixed to
this DEED in the presence of: [COMMON SEAL]

(Signed) JOHN DOE Director

(Signed) JANE JONES Secretary

A company not using a common seal could execute as follows:

SIGNED as a DEED for
and on behalf of
JOHN DOE [LIMITED] OR
[PLC] by:

(Signed) JOHN DOE Director

(Signed) JANE JONES Secretary

Attestation clauses should preferably be added at the very end of the relevant document, i.e., *after* any annexations or schedules.

If, exceptionally, a document is executed otherwise than as a Deed, all references to Deed in the relevant text should be replaced by Agreement , and the attestation clauses should be as follows:

SIGNED by JOHN DOE
in the presence of:

OR

SIGNED by JOHN DOE
for and on behalf of
JOHN DOE [LIMITED]
OR [PLC] in the presence of:

Execution of self-proving and other documents under Scots law

Under Scots law, a self-proving document has the advantage of being presumed to have been validly signed, and it is therefore recommended that the formalities of witnessing be followed.

An individual or sole trader should sign his usual signature in the presence of one witness who must also sign, and whose full name and address must appear in the testing clause (the Scottish terminology for an attestation clause). An appropriate testing clause is as follows:

SIGNED by JOHN DOE at   on the   day of   before the undernoted witness:

(Signed) RICHARD ROE   (Signed) JOHN DOE

of:

In the case of a partnership, the document should be signed on behalf of the partnership by one of the partners, by signing either the firm name or his own name. In all cases, to be self-proving the signature should be in the presence of one witness, who must also sign and whose full name and address must appear in the testing clause. This may be subject to alternative methods of execution provided for in the partnership agreement. A suggested testing clause for a partnership is as follows:

SIGNED for and on behalf of the JOHN DOE PARTNERSHIP at   on the   day of   by JOHN DOE, one of the partners thereof] OR [the firm name being adhibited by JOHN DOE, one of the partners thereof], before the undernoted witness:

(Signed) RICHARD ROE   (Signed) [JOHN DOE]
                                                                OR
                                                [JOHN DOE PARTNERSHIP] of:

A company executing a document under Scots law does not require to affix its seal. Although execution by two directors, or one director and the company secretary, or by two authorised persons (as provided for in the Companies Act 1985) is still a competent method of execution, the method introduced by the Requirements of Writing (Scotland) Act 1995 is the signature of one director or the company secretary or an authorised person before one witness whose full name and address is stated in the testing clause. If the document is signed by an authorised person, evidence of such authority should be produced. An appropriate testing clause for a company would be as follows:

SIGNED for and on behalf of JOHN DOE [LIMITED] OR [PLC] by JOHN DOE [a director] OR [its secretary] OR [a person authorised to sign on its behalf], at   on the   day of   before the undernoted witness:

(Signed) RICHARD ROE   (Signed) JOHN DOE

of:

The testing clause for at least one signatory must appear on the same page as at least part of the body or operative part of the document in question - that is, the part of the document before 'IN WITNESS whereof...' The testing clauses must appear at the end of the principal document, i.e. before any annexations or schedules.

There are particular rules under Scots law regarding annexations, which must be followed carefully to ensure that the annexation is validly incorporated. For an annexation to be validly incorporated it must be referred to in the document and identified on its face as being the annexation referred to in the document. If the annexation is a drawing, plan, photograph or other such representation it should be signed on every page by each party. If the annexation is another form of writing, such as an inventory, appendix or schedule, which describes or shows any part of the land to which the Contract relates, it need only be signed on the last page. Witnesses are not required to sign annexations. To avoid any doubt, it is recommended that all annexations (other than those used which require signing on each page) be signed on the last page.

The actual date of the signing by each party of the document is inserted in the testing clause relating to that party. Thus, unlike English practice, a document may state a number of different signing dates. In preparing for execution any ancillary document to the Contract, such as a performance bond or parent company guarantee, where Scots law applies the reference at the beginning of the document to its being made on a certain date should be deleted. In each case, the words at the end following 'IN WITNESS whereof' should be deleted, and substituted by the words 'these presents typewritten on this and the preceding [ ] pages(s) are executed as follows:'. Testing clauses as described above for each signatory should then follow. It should also be noted that under Scots law an unilateral obligation does not require consideration for its validity.

Execution of documents under foreign law

If a document is to be executed under foreign law - that is, a system of law other than English, Scots or Northern Ireland law - specific professional advice should be taken in connection with the relevant jurisdiction. This issue will arise, for example, if the Contractor or other parties, such as parent company guarantors, are incorporated under foreign law.

Multi-party disputes and the Arbitration Act 1996

The Arbitration Act 1996 is intended to restate and improve the law relating to arbitration. It extends to England and Wales and, in all material respects, to Northern Ireland. However, it only extends to Scotland in certain immaterial respects.

The Act has resulted in a significant change in arbitration law and procedure, by effectively narrowing the grounds upon which the courts may disregard arbitration agreements, and adjudicate relevant disputes themselves. This is explained below.

Under Section 9(1) of the Act, a party to an arbitration agreement against whom legal (court) proceedings are brought in respect of a matter which, under the agreement, is to be referred to arbitration, may apply to the court to stay (stop) those proceedings. Under Section 9(4), on such an application 'the court shall grant a stay unless satisfied that the arbitration agreement is null and void, inoperative, or incapable of being performed'. Therefore, the party bringing the legal proceedings will be compelled to arbitrate, unless he can establish the narrow exceptions stated in Section 9(4).

Under Section 86(1) of the Act, Section 9(4) 'does not apply to a domestic arbitration agreement'. That term defined by Section 85(2) as an arbitration agreement to which none of the parties is -

(a) an individual who is a national of, or habitually resident in, a state other than the United Kingdom, or

(b) a body corporate which is incorporated in, or whose central control and management is exercised in, a state other than the United Kingdom

and under which the seat of the arbitration (if the seat has been designated or determined) is in the United Kingdom.

Section 86(2) provides that, on an application under Section 9 to stay legal proceedings 'in relation to a domestic arbitration agreement the court shall grant a stay unless satisfied -

(a) that the arbitration agreement is null and void, inoperative, or incapable of being performed, or

(b) that there are other sufficient grounds for not requiring the parties to abide by the arbitration agreement'.

Therefore, in relation to domestic arbitration agreements, Section 86(2)(b) was obviously intended to give the court a much wider discretion to refuse a stay than Section 9(4). However, the Arbitration Act 1996 (Commencement No. 1) Order 1996, while it brought the rest of the Act into force on 31 January 1997, did not bring into force Sections 85 to 87, which make the above and other special provisions in relation to 'domestic arbitration agreements'. Therefore, Section 9(4) is in force, but Section 86 is not, with the result that the wider discretion of the court under Section 86(2)(b) is inoperative.

This apparently obscure point has considerable practical importance in relation to construction arbitration, where multi-party disputes are very common. Under the inoperative Section 86(2)(b), and under the previous law, the courts could refuse a stay if satisfied that there were 'other' sufficient grounds for not requiring the parties to abide by the arbitration agreement. One such ground could be that the dispute involved several parties under different contracts - for example, if a construction defect exists, the causation of, and responsibility for, that defect may be the subject of disputes between the Employer, the main Contractor, his subcontractors, the Employer's architect and the Employer's consulting engineers. Unlike arbitrators, the courts have compulsory powers under the general law to consolidate the legal proceedings in such disputes, and to order concurrent hearings. However, the courts no longer have power to refuse to stay legal proceedings on such grounds. The only grounds for refusing a stay are stated in Section 9(4). It should be noted that in Scotland there is no statutory provision giving the courts the right to refuse a sist (the equivalent of a stay) for arbitration, and the courts discretion to refuse a sist is limited. It would not normally be sufficient grounds for refusing a sist that the dispute involved other parties under different contracts.

Therefore, in multi-party disputes arising out of the same facts, different tribunals, whether arbitrators or courts, may reach different decisions. For example, an arbitrator between the Employer and the main Contractor may decide that a defect in the Works is due to a defect in the Employer's architect's design, while a court or arbitrator adjudicating on a parallel dispute between the Employer and his architect may decide that the defect is due to a defect in the main Contractor's materials and workmanship. Therefore, the Employer could receive nothing, although he had received, and paid for, defective Works.

The Employer may protect himself from such problems by ensuring:

- that all his contracts with contractors, consultants and others relating to the same project provide for the arbitration of disputes;

- that the same arbitrator is appointed under all the contracts, by naming him in each; and

- that the arbitrator has power to order that arbitral proceedings under each of the contracts shall be 'consolidated' (proceed together) and that concurrent hearings shall be held (as contemplated by Section 35 of the Arbitration Act 1996) and, in Scotland, that the arbiter has the same power as the court in respect of the joining of one or more defenders or joining co-defenders or third parties.

These are the solutions adopted in the *GC/Works* arbitration provisions.

Scope of the Housing Grants, Construction and Regeneration Act 1996

While the *GC/Works (1998)* forms of contract have been drafted with the intention of fully complying with Part II (Construction contracts) of the Act (and with the Construction Contracts (Northern Ireland) Order 1997), it should be appreciated that not all construction contracts are covered by the new legislation. This is because of the specialised statutory definitions of the construction contracts and construction operations to which the legislation applies.

Section 104(1) of the Act defines 'construction contract' as (inter alia) 'an agreement with a person for...the carrying out of construction operations' , and Section 105(1) defines 'construction operations' as:

'operations of any of the following descriptions -

(a) construction, alteration, repair, maintenance, extension, demolition or dismantling of buildings, or structures forming, or to form, part of the land (whether permanent or not); (b) construction, alteration, repair, maintenance, extension, demolition or dismantling of any works forming, or to form, part of the land, including (without prejudice to the foregoing) walls, roadworks, power-lines, telecommunication apparatus, aircraft runways, docks and harbours, railways, inland waterways, pipe-lines, reservoirs, water-mains, wells, sewers, industrial plant and installations for purposes of land drainage, coast protection or defence;

(c) installation in any building or structure of fittings forming part of the land, including (without prejudice to the foregoing) systems of heating, lighting, air-conditioning, ventilation, power supply, drainage, sanitation, water supply or fire protection, or security or communication systems;

(d) external or internal cleaning of buildings and structures, so far as carried out in the course of their construction, alteration, repair, extension or restoration;

(e) operations which form an integral part of, or are preparatory to, or are for rendering complete, such operations as are previously described in this subsection, including site clearance, earth- moving, excavation, tunnelling and boring, laying of foundations, erection, maintenance or dismantling of scaffolding, site restoration, landscaping and the provision of roadways and other access works;

(f) painting or decorating the internal or external surfaces of any building or structure'.

The definition will cover most forms of non-domestic construction; alteration, repair, maintenance, extension, demolition or dismantling of buildings; civil, mechanical and electrical engineering works; and ancillary or preparatory works. The definition is wide, extending, for example, under Section 105(1)(f), to painting or decorating internal or external surfaces.

However, Section 105(2) makes important exceptions, by providing that:

'The following operations are not construction operations within the meaning of this Part -

(a) drilling for, or extraction of, oil or natural gas;

(b) extraction (whether by underground or surface working) of minerals; tunnelling or boring, or construction of underground works, for this purpose;

(c) assembly, installation or demolition of plant or machinery, or erection or demolition of steelwork for the purposes of supporting or providing access to plant or machinery, on a site where the primary activity is -

  (i)   nuclear processing, power generation, or water or effluent treatment, or
  (ii) the production, transmission, processing or bulk storage (other than warehousing) of chemicals, pharmaceuticals, oil, gas, steel or food and drink;

(d) manufacture or delivery to site of -

  (i)   building or engineering components or equipment,
  (ii) materials, plant or machinery, or
  (iii) components for systems of heating, lighting, air-conditioning, ventilation, power supply, drainage, sanitation, water supply or fire protection, or for security or communications systems,

  except under a contract which also provides for their installation;

(e) the making, installation and repair of artistic works, being sculptures, murals and other works which are wholly artistic in nature'.

While, no doubt, it will be unusual for Crown Employers to enter into construction contracts within the above exceptions, it is possible that they may do so. Paragraph (c), in particular, could cover quite a wide range of projects in relation to the industrial activities mentioned.

There are other exceptions: under Section 106, the Act does not apply to construction contracts with residential occupiers for operations on their houses and flats, and under Section 107, the Act does not apply to construction contracts which are not in writing. However, neither of these exceptions has any practical relevance to Crown Employers.

The Secretary of State has power under Section 104(4) to amend the 'construction contract' definition: under Section 105(3) to amend the construction operations definition: and, under Section 106, to exclude classes of construction contract from the operation of the Act.

The *GC/Works (1998)* forms of contract may be used unamended in relation to construction contracts which are outside the scope of the Act. However, the forms contain certain Conditions which have been inserted solely in order to comply with the Act. In *GC/Works/2 (1998)*, these are Conditions 33 (Withholding payment) and 35 (Suspension for non-payment), which respectively reflect Sections 111 and 112 of the Act.

If the Contract is outside the scope of the Act, and the Employer after due consideration concludes that it is appropriate to delete those Conditions, and the cross-references to those Conditions in other Conditions, this may be effected by a Supplementary Condition, upon which specific legal advice should be taken.

Certain other Conditions are also intended to comply with the Act - for example, Condition 42 (Adjudication) reflects Section 108, Conditions 30 (Advances on account), 31 (Final Account) and 32 (Certifying payments) reflect Sections 109 and 110, and Conditions 1(3) and (4) (Definitions, etc.) reflect Section 115 and 116. However, it will not be necessary or appropriate to amend those Conditions, solely on the grounds that the Contract may be outside the scope of the Act.

GC/WORKS/2 EDITION 2 1990
DESTINATION TABLE

GC/WORKS/2 EDITION 2 1990			GC/WORKS/2(1998)
CONDITION	**NO.**	**NO.**	**CONDITION**
Definitions etc.	1	1	Definitions, etc.
Definition of 'Contract'	1(1)	1(1)	
Definitions of:			
'Accepted Risks'	1(2)	1(1)	
'Authority'	1(2)	1(1)	'Employer'
'Contract Sum'	1(2)	1(1)	
'Contractor'	1(2)	1(1)	
'Date for Completion'	1(2)	1(1)	
'Final Sum'	1(2)	1(1)	
'Site'	1(2)	1(1)	
'SO' or 'Superintending Officer'	1(2)	1(1)	'PM' or 'Project Manager'
'Unforeseen Ground Conditions'	1(2)	1(1)	Unforeseeable Ground Conditions'
'Works'	1(2)	1(1)	
Definitions of 'Things' 'Things' for incorporation and 'Things' not for incorporation	1(3)	1(1)	
Employer's decisions	1(4)	3	Delegations and representatives
Interpretation	1(5)	1(2)	
Notices	1(6)	1(3)	
Contractor deemed to have satisfied himself as to conditions affecting execution of the Works	2	4	Conditions affecting Works
Unforeseeable Ground Conditions	2A	4	Conditions affecting Works
Vesting of Works etc. in the Authority. Things not to be removed	3	18	Vesting
Specifications and Drawings	4	2	Contract documents
Progress of the Works	5	21	Commencement and completion

CONDITION	NO.	NO.	CONDITION
SO's instructions	6	25	PM's Instructions
Valuation of the SO's instructions	7	26	Valuation of Instructions
Things for incorporation and workmanship to conform to description	8	19	Quality
Local and other authorities' notices and fees	9	6	Statutory notices and CDM Regulations
Watching, lighting and protection of Works	10	7(1)	Protection of Works
Removal of rubbish	11	21(4)	Commencement and completion
Daywork	12	13(3)	Records
Precautions against fire and other risks	13	7	Protection of Works
Damage to Works or other things	14	8	Loss or damage
Assignment or transfer of Contract	15	44	Assignment and subletting
Date for completion: Extensions of time	16	21	Commencement and completion
		23	Extensions of time
Sub-letting	17	44	Assignment and subletting
Sub-contractors and suppliers	18	44	Assignment and subletting
Defects liability	19	9	Defects in Maintenance Periods
Contractor to conform to regulations	20	10	Occupier's rules and regulations
Prime Cost items	21	-	No equivalent
Provisional sums	22	45	Provisional sums
Advances on account	23	30	Advances on account
Payment on and after completion	24	31	Final Account
Recovery of sums due from the Contractor	25	34	Recovery of sums
Determination of Contract due to default or failure of Contractor	26	38	Determination by Employer
Provisions in case of determination of Contract	27	39	Consequences of determination by Employer
Injury to persons: Loss of property	28	8	Loss or damage
Facilities for other works	29	46	Other works
(Not used)	30	-	-

CONDITION	NO.	NO.	CONDITION
Racial discrimination	31	11	Discrimination
Corrupt gifts and payment of commission	32	12	Corruption
Admission to the Site	33	14	Site admittance
Passes	34	15	Passes
Photographs	35	16	Photographs
Secrecy	36	17	Official secrets and confidentiality
Arbitration	37	43	Arbitration and choice of law

GC/WORKS/2 (1998) DERIVATION TABLE

GC/WORKS/2 (1998)			GC/WORKS/2 EDITION 2 1990
CONDITION	**NO.**	**NO.**	**CONDITION**
Definitions, etc.	1	1	Definitions etc.
Definitions of:			
'Abstract of Particulars'	1(1)	1(1)	
'Accepted Risks'	1(1)	1(2)	
'CDM Regulations'	1(1)	-	No equivalent
'Conditions of Contract'	1(1)	-	No equivalent
'Contract'	1(1)	1(1)	
'Contract Sum'	1(1)	1(2)	
'Contractor'	1(1)	1(2)	
'Crown'	1(1)	-	No equivalent
'Date for Completion'	1(1)	1(2)	
'Days'	1(1)	-	No equivalent
'Employer'	1(1)	1(2)	'Authority'
'Final Account'	1(1)	-	No equivalent
'Final Sum'	1(1)	1(2)	
'Health and Safety Plan'	1(1)	-	No equivalent
'Instruction'	1(1)	6	'SO's instructions'
'Maintenance Period'	1(1)	19	Defects liability
'Order to Proceed'	1(1)	-	No equivalent
'Planning Supervisor'	1(1)	-	No equivalent
'Principal Contractor'	1(1)	-	No equivalent
'PM'	1(1)	1(2)	'SO'
'Site'	1(1)	1(2)	
'Specification' and/or 'Drawings'	1(1)	1(1)	

CONDITION	NO.	NO.	CONDITION
'Things' , 'Things for incorporation and' 'Things not for incorporation'	1(1)	1(3)	
'Unforeseeable Ground Conditions'	1(1)	1(2)	'Unforeseen Ground Conditions'
'Works'	1(1)	1(2)	
Interpretation	1(2)	1(5)	
Notices	1(3)	1(6)	
Periods of time	1(4)	-	No equivalent
Fair dealing	1A	-	No equivalent
Contract documents	2	4	Specifications and Drawings
Delegations and representatives	3	1(4)	
Conditions affecting Works	4	2	Contractor deemed to have satisfied himself as to conditions affecting execution of the Works
		2A	Unforeseeable Ground Conditions
Insurance	5	-	No equivalent
Statutory notices and CDM Regulations	6	9	Local and other authorities' notices and fees
Protection of Works	7	10	Watching, lighting and protection of Works
		13	Precautions against fire and other risks
Loss or damage	8	14	Damage to Works or other things
Defects in Maintenance Periods	9	19	Defects liability
Occupier's rules and regulations	10	20	Contractor to conform to regulations
Discrimination	11	31	Racial discrimination
Corruption	12	32	Corrupt gifts and payment of commission
Records	13	12	Daywork
Site admittance	14	33	Admission to the Site
Passes	15	34	Passes

CONDITION	NO.	NO.	CONDITION
Photographs	16	35	Photographs
Official secrets and confidentiality	17	36	Secrecy
Vesting	18	3	Vesting of Works etc. in the Authority. Things not to be removed
Quality	19	8	Things for incorporation and workmanship to conform to description
Excavations	20	-	No equivalent
Commencement and completion	21	5	Progress of the Works
		11	Removal of rubbish
		16	Date for completion: Extensions of time
Progress meetings	22	-	No equivalent
Extensions of time	23	16	Date for completion: Extensions of time
Certifying completion	24	-	No equivalent
PM's Instructions	25	6	SO's instructions
Valuation of Instructions	26	7	Valuation of the SO's instructions
VAT	27	-	No equivalent
Prolongation and disruption	28	-	No equivalent
Finance charges	29	-	No equivalent
Advances on account	30	23	Advances on account
Final Account	31	24	Payment on and after completion
Certifying payments	32	-	No equivalent
Withholding payment	33	-	No equivalent
Recovery of sums	34	25	Recovery of sums due from the Contractor
Suspension for non-payment	35	-	No equivalent
Non-compliance with Instructions	36	-	No equivalent
Damages for delay	37	-	No equivalent

CONDITION	NO.	NO.	CONDITION
Determination by Employer	38	26	Determination of Contract due to default or failure of Contractor
Consequences of determination by Employer	39	27	Provisions in case of determination of Contract
Determination by Contractor	40	-	No equivalent
Determination following suspension of Works	41	-	No equivalent
Adjudication	42	-	No equivalent
Arbitration and choice of law	43	37	Arbitration
Assignment and subletting	44	15	Assignment or transfer of Contract
		17	Sub-letting
		18	Sub-contractors and suppliers
Provisional sums	45	22	Provisional sums
Other works	46	29	Facilities for other Works
Performance bond	47	-	No equivalent
Parent company guarantee	48	-	No equivalent

ACKNOWLEDGEMENTS

The Government agency responsible for this document is -

Property Advisers to the Civil Estate (PACE)
Trevelyan House
Great Peter Street
London SW1P 2BY

The development of this document was directed and greatly aided by a sub-group of the PACE Joint Users' Group (JUG) of Government Departments, consisting of -

Bruce Perry	PACE Central Advice Unit (Chairman)
Robert Boyd	Department of the Environment for Northern Ireland Construction Service
Charles Branch	Department of Trade and Industry
Ian Campbell	The Scottish Office
Mike Frazer	Benefits Agency Estates
John Garnett	Consultant to the Ministry of Agriculture, Fisheries and Food
Jeff Hogg	Lord Chancellor's Department
Jay Jayasundara	HM Treasury (Procurement Practice & Development Group, formerly Central Unit on Procurement)
Lynne Jones	The Buying Agency
Graham Mason	Prison Service
Deric McTier	Department of the Environment for Northern Ireland Construction Service
Robert Pilling	PACE Central Advice Unit
David Reid	Home Office
Bill Robinson	Employment Service
Richard Whittaker	English Heritage
Les Wilson	British Nuclear Fuels plc
David Woolger	Inland Revenue
Neil Wright	Benefits Agency Estates
Harry Yeabsley	Ministry of Defence

The substantive drafting and legal advice and assistance necessary for the preparation of this document was principally provided by Andrew Pike of -

Pinsent Curtis
Solicitors 3
Colmore Circus
Birmingham B4 6BH

Advice on Scots law was provided by David Henderson of -

MacRoberts
Solicitors
152 Bath Street
Glasgow G2 4TB

Advice on Northern Ireland law was provided by -

Departmental Solicitor's Office
Department of Finance & Personnel
Victoria Hall
12 May Street
Belfast BT1 4NL

Advice on risk and insurance matters was provided by Robin Keeling of -

Aon Group Limited
Construction Division
15 Minories
London EC3N 1NJ

Construction consultancy advice was provided by Brendan Murphy of -

Tarmac Services
The Lansdowne Building
Lansdowne Road
Croydon CR0 2BX

Printed in the United Kingdom for The Stationery Office
J0042679 3/98 C20 10170